AGRICULTURE ISSUES AND POLICIES

U.S. GRAIN CONSUMPTION

AGRICULTURE ISSUES AND POLICIES

Additional books in this series can be found on Nova's website under the Series tab.

Additional E-books in this series can be found on Nova's website under the E-books tab.

AGRICULTURE ISSUES AND POLICIES

U.S. GRAIN CONSUMPTION

SARA D. TORRES
AND
DANIEL M. VARGAS
EDITORS

Nova Science Publishers, Inc.
New York

Copyright © 2011 by Nova Science Publishers, Inc.

All rights reserved. No part of this book may be reproduced, stored in a retrieval system or transmitted in any form or by any means: electronic, electrostatic, magnetic, tape, mechanical photocopying, recording or otherwise without the written permission of the Publisher.

For permission to use material from this book please contact us:
Telephone 631-231-7269; Fax 631-231-8175
Web Site: http://www.novapublishers.com

NOTICE TO THE READER

The Publisher has taken reasonable care in the preparation of this book, but makes no expressed or implied warranty of any kind and assumes no responsibility for any errors or omissions. No liability is assumed for incidental or consequential damages in connection with or arising out of information contained in this book. The Publisher shall not be liable for any special, consequential, or exemplary damages resulting, in whole or in part, from the readers' use of, or reliance upon, this material. Any parts of this book based on government reports are so indicated and copyright is claimed for those parts to the extent applicable to compilations of such works.

Independent verification should be sought for any data, advice or recommendations contained in this book. In addition, no responsibility is assumed by the publisher for any injury and/or damage to persons or property arising from any methods, products, instructions, ideas or otherwise contained in this publication.

This publication is designed to provide accurate and authoritative information with regard to the subject matter covered herein. It is sold with the clear understanding that the Publisher is not engaged in rendering legal or any other professional services. If legal or any other expert assistance is required, the services of a competent person should be sought. FROM A DECLARATION OF PARTICIPANTS JOINTLY ADOPTED BY A COMMITTEE OF THE AMERICAN BAR ASSOCIATION AND A COMMITTEE OF PUBLISHERS.

Additional color graphics may be available in the e-book version of this book.

LIBRARY OF CONGRESS CATALOGING-IN-PUBLICATION DATA
U.S. grain consumption / editors: Sara D. Torres and Daniel M. Vargas.
p. cm. -- (Agriculture issues and policies series)
 Includes bibliographical references and index.
 ISBN 978-1-61122-953-0 (softcover : alk. paper)
 1. Grain--United States. 2. Grain trade--United States. 3. Food consumption--United States--Statistics. 4. Specialty crops--Economic aspects--United States. I. Torres, Sara D. II. Vargas, Daniel M. III. Series: Agriculture issues and policies series.
 HD9035.U232 2011
 339.4'863310973--dc22
 2010047108

Published by Nova Science Publishers, Inc. † New York

CONTENTS

Preface		vii
Chapter 1	The U.S. Grain Consumption Landscape: Who Eats Grain, in What Form, Where and How Much? *Bing-Hwan Lin and Steven T. Yen*	1
Chapter 2	The Changing Face of the U.S. Grain System: Differentiation and Identity Preservation Trends *Aziz Elbehri*	37
Index		81

PREFACE

Dietary Guidelines, issued by the U.S. Department of Health and Human Services and the U.S. Department of Agriculture, are intended to help consumers choose diets that meet their nutritional needs and improve their health. As part of a healthy diet, the Guidelines emphasize the value of whole grains. There is growing evidence that those who consume enough whole grains may reduce their risk of heart disease as well as their likelihood of becoming overweight. This book examines grain consumption by economic and demographic characteristics of consumers, as well as the effects of consumer's social, economic and demographic characteristics and dietary perceptions and practices.

Chapter 1- The 2005 Dietary Guidelines, issued by the U.S. Department of Health and Human Services and the U.S. Department of Agriculture, are intended to help consumers choose diets that meet their nutritional needs and improve their health. As part of a healthy diet, the *Guidelines* emphasize the value of whole grains. There is growing evidence that those who consume enough whole grains may reduce their risk of heart disease as well as their likelihood of becoming overweight.

Chapter 2- The U.S. grain system is increasingly marked by product differentiation and market segmentation. More specialty crops now require either some form of segregation or full-scale identity preservation to keep them separate from conventional commodities. Market segmentation within the grain system is driven by the need to preserve its market value, or ensure purity of the product. Internationally, U.S. grain markets must increasingly conform to a new regulatory environment reliant on traceability and identity preservation.

In: U.S. Grain Consumption
Editors: Sara D. Torres et al.

ISBN: 978-1-61122-953-0
© 2011 Nova Science Publishers, Inc.

Chapter 1

THE U.S. GRAIN CONSUMPTION LANDSCAPE: WHO EATS GRAIN, IN WHAT FORM, WHERE AND HOW MUCH?*

Bing-Hwan Lin and Steven T. Yen

ABSTRACT

The U.S. Government is promoting whole-grain foods, responding to mounting evidence of their association with maintaining a healthy weight and reducing the risk of heart problems and other diseases. This study compared Americans' consumption of grains with the recommendations in the Government's *2005 Dietary Guidelines*, using data from USDA's *Continuing Survey of Food Intakes by Individuals*, 1994-96 and 1998. The analysis confirmed a national preference for refined grains—only 7 percent of survey respondents met the 2005 whole-grain recommendation. The authors compared grain consumption by economic and demographic characteristics of consumers, and also examined the effects of consumers' social, economic, and demographic characteristics and dietary perceptions and practices. The results suggest that consumers who perceive grain consumption as important and read food labels during shopping tend to eat more whole grains than other people. When data from more recent surveys are analyzed, results of the present study can

---

* This is an edited, reformatted and augmented edition of a United States Department of Agriculture Economic Research Report Number 50, dated November 2007.

serve as a baseline from which to gauge changes in the American diet and the consumption of whole grains.

Keywords: Whole grain consumption, grain consumption, dietary guidelines, food consumption survey data.

ACKNOWLEDGMENTS

The authors appreciate the comments and suggestions from Jay Variyam and Elise Golan of the USDA Economic Research Service, Chung L. Huang of the University of Georgia, and the three anonymous reviewers. Special thanks go to Courtney Knauth for editorial assistance and to Susan DeGeorge for layout and design.

SUMMARY

The *2005 Dietary Guidelines*, issued by the U.S. Department of Health and Human Services and the U.S. Department of Agriculture, are intended to help consumers choose diets that meet their nutritional needs and improve their health. As part of a healthy diet, the *Guidelines* emphasize the value of whole grains. There is growing evidence that those who consume enough whole grains may reduce their risk of heart disease as well as their likelihood of becoming overweight.

What Is the Issue?

Are Americans actually following the grain consumption recommendations in the *2005 Dietary Guidelines?* More specifically, how much grain do Americans eat? At which meals? What characteristics are associated with low or high consumption of refined and whole grains? Which subpopulations are particularly deficient in meeting the whole-grain recommendations? Answers to these questions can serve as guidelines for developing intervention strategies.

What Did the Study Find?

The analysis showed a strong preference in the American diet for refined grains over whole grains. Ninety-three percent of Americans failed to meet the recommendation to consume 3 ounces per day of whole grains for a 2,000-calorie diet. Specific findings include:

- Americans eat too much refined grain and not enough whole grain. During 1994-96 and 1998, Americans consumed 6.7 ounces of total grains per day, or 106 percent of the recommendation. However, they overconsumed refined grains, averaging 77 percent more than the recommended daily amount, while eating 34 percent of the amount of whole grains recommended in the *2005 Dietary Guidelines*. Children, even more than adults, favored refined over whole grains, and the presence of children in the home had a negative effect on adults' whole-grain consumption.
- Breakfast foods are good sources of whole grains. Americans ate 40 percent of their whole grains at breakfast, 23 percent at lunch, and 17 percent at dinner, with the rest provided by snack foods.
- Restaurant foods are not a good source of whole grain. A third of Americans' calories came from meals prepared away from home, yet 1,000 calories of a restaurant meal averages less than one-third ounce of whole grains. Thus, it takes over 10,000 calories of restaurant food to obtain the amount of whole grains needed to meet the Government guidelines.
- Grain consumption varies by race and ethnicity. The study found that Asians averaged 22 percent of the recommended amount of whole grains, compared with 25 percent for Blacks, 35 percent for Whites, and 41 percent for Hispanics.
- Food-label use matters, as do personal perceptions about grains in the diet. Both food-label use (or non-use) and an individual's perception of whether grains affect health influenced the person's total grain intake, with perception having the greater impact. Those who considered it important to eat enough grains were 36 percent more likely to consume whole grains than those who did not.
- Some demographic characteristics are associated with grain consumption. Individuals most likely to read food labels and to value grains in the diet included those with higher educational attainment, meal planners, and people who exercise vigorously. Higher household income

was associated with the use of food labels, but not with the perceived importance of grain consumption. People less likely to use food labels and to consider grains important included smokers and those who doubted that food choices affected health.

How Was the Study Conducted?

The authors analyzed data from USDA's *Continuing Survey of Food Intakes by Individuals* (CSFII) conducted in 1994-96 and 1998. The survey also collected various economic, social, and demographic characteristics for each respondent and his/her household. The 1994-96 survey had a companion module, *The Diet and Health Knowledge Survey*, which asked adults about their information, attitudes, and practices with respect to diet and health, making the CSFII data ideal for examining the effects of knowledge and practices on food consumption. Since 1998, USDA has published two further surveys of U.S. food intake, most recently for 2003-2004. However, these surveys did not ask about dietary knowledge and practices and cannot be used to study their effects on grain consumption. When data from future surveys are analyzed, the present study will be valuable as a baseline for assessing changes in the U.S. diet and the consumption of grains and whole grains.

INTRODUCTION

Grain products are available in two basic forms, refined and whole. Whole grains contain all three key parts of the kernel—the bran, the germ, and the endosperm. Refining normally removes most of the bran and some of the germ. Grains can be labeled as whole grains if they contain the same proportions of bran, germ, and endosperm as in the original grain. Whole-grain products are noticeably darker than refined white products due to the presence of bran. Historically, there has been a belief that white flour was the food of the rich and unrefined flour the food of the hard-working peasant and the poor (Spiller, 2002). Americans tend to favor—by a substantial margin— refined grains over whole grains. The box "Sources for Information About Grains" (p.2) is for readers who are interested in learning more about refined and whole grains.

Nutritional Superiority of Whole Grains

Even though some vitamins and minerals are added back to enrich refined grains, whole grains provide greater amounts of vitamins, minerals, fiber, and other valuable substances. Responding to mounting evidence of the association between whole-grain consumption and a reduced risk of heart problems and other diseases, as well as an association with healthy weight maintenance, the U.S. Government and the food industry have been promoting grains— especially whole grains— in the American diet.

In a comprehensive review of scientific evidence, the National Research Council (1989) concluded that "Diets high in plant foods—i.e., fruits, vegetables, legumes, and whole-grain cereals—are associated with a lower occurrence of coronary heart disease and cancers of the lung, colon, esophagus, and stomach." This scientific consensus, together with subsequent research on whole-grain foods, was the basis for the whole-grains/cancer-and-heart disease health claim submitted by General Mills for its Cheerios cereals in 1999, and approved by the Food and Drug Administration (Wiemer, 2002).

The national goals specified in *Healthy People 2010* include the objective of increasing the proportion of people who consume at least six daily servings of grain products, with at least three servings of whole grains (USDHHS, 2005). During 1994-96, only half of Americans ate 6 or more servings of grain products a day, and only 1 in 10 ate 3 or more servings of whole-grain products a day (Kantor et al., 2001). The *2005 Dietary Guidelines* made the first recommendation for a specific number of whole-grain servings by caloric intake (USDA and USDHHS, 2005). The grain industry and the public health community share an interest in increasing whole-grain consumption, through both marketing strategies and public health campaigns.

Understanding U.S. Grain Consumption

Who consumes grain products, and how much do they consume? Where and at what meal occasions do Americans consume grains? What are the factors associated with low or high consumption of refined and whole grains? Which subpopulations fall particularly short of the recommendations? This study sought answers to such information, which has been very limited (Harnack et al., 2003; Kantor et al., 2001; Moutou et al., 1998).

The objectives of the study were twofold:

- To describe U.S. consumption of refined and whole grains and compare the amounts consumed against the *2005 Dietary Guidelines* recommendation by economic and demographic characteristics of consumers. This descriptive information points out a general dietary deficiency in grain consumption, paving the way for the development of intervention strategies. The descriptive statistics can serve as the baseline for monitoring national progress in meeting the Federal recommendations for whole- grain consumption.
- To conduct a regression analysis to identify social, economic, demographic, nutrition knowledge, and behavioral factors associated with consumption of refined- and whole-grain products.

To accomplish these objectives, we analyzed data from the USDA's *Continuing Survey of Food Intakes by Individuals* (CSFII) conducted in 1994-96 and 1998 (USDA, 2000). The 1994-96 survey included a companion module that asked adults about their knowledge, attitudes, and practices with respect to diet and health, making the CSFII data ideal for examining the formation of dietary knowledge/attitudes, the adoption of dietary practices, and the effects of knowledge and practices on food consumption. Even though grains are a staple in the American diet, many consumers do not eat whole grains regularly. This data characteristic—zero consumption—complicates the econometric modeling.

SOURCES FOR INFORMATION ABOUT GRAINS

Information on grains can be obtained from the Federal Government, the grain and food industry, and not-for-profit organizations. The Federal Government provides accurate information to help Americans choose healthy diets at www.HealthierUS.gov, in the section devoted to nutrition. The book *Whole-Grain Foods in Health and Disease* (Marquart et al., 2002) gives a comprehensive review of current whole- grain science and technology, regulatory and policy issues, dietary intake, consumer interest, and health promotion. The Whole Grains Council has a consumer guide on the benefits of whole grains (*www.wholegrainscouncil.org/consumer%20 guide.html*). Many other not-for-profit organizations, including universities and medical schools, provide valuable information on whole grains.

USDA FOOD CONSUMPTION SURVEY DATA

USDA has conducted periodic food consumption surveys in the United States since the 1930s. Data from the 1994-96 and 1998 *Continuing Survey of Food Intakes by Individuals* (CSFII) and the 1994-96 *Diet and Health Knowledge Survey* (DHKS) are analyzed in this study. A major task of the CSFII was to collect data on dietary intakes. Similar intake data were also collected in the *National Health and Nutrition Examination Survey* (NHANES) conducted by the National Center of Health Statistics of the Centers for Disease Control and Prevention (CDC). CSFII and NHANES surveys have been integrated since 2002, at which time the DHKS survey was dropped. Therefore, the 1994-96 CSFII and DHKS surveys provide the only national data for examining the relationship between dietary knowledge and attitudes and dietary/health outcomes.

The CSFII/DHKS survey was first implemented in 1989-91, and was repeated in 1994-96. Each year of the 1994-96 CSFII survey comprises a nationally representative sample of noninstitutionalized persons residing in the United States. The 1998 CSFII is a supplemental survey to increase the sample of children in the 1994-96 CSFII. The DHKS surveyed only adults, and hence was not conducted in 1998.

In the CSFII, 2 nonconsecutive days of dietary data were collected for individuals of all ages 3 to 10 days apart, through in-person interviews using 24-hour recall. The 1994-96 CSFII data provide information on the food intakes of 15,303 individuals, who gave a list of foods consumed, where the food was prepared and eaten, how much was eaten, and at what meal and time. After the respondents reported their first day of dietary intake, an adult 20 years or older was randomly selected from each household to participate in the DHKS. The DHKS questions cover a wide range of topics, including self-perception of the adequacy of nutrient intakes, awareness of diet-health relationships, knowledge of dietary recommendations, perceived importance of following dietary guidelines, use and perception of food labels, and behavior related to fat intake and food safety. Of the 7,842 households eligible to participate in the DHKS, respondents from 5,765 households completed the survey. The 1998 CSFII survey collected dietary data for 5,559 children up to age 9. Because only children were surveyed, the DHKS was not conducted in 1998. Various economic, social, and demographic characteristics were also collected for the CSFII respondent and his/her household.

The Agricultural Research Service (ARS) of USDA has created several technical databases, including the Pyramid Servings Database (PSD), to support use of the CSFII data. For 30 food groups, including refined and whole grains, the PSD converts the amount of food consumed into the number of servings, enabling comparison with dietary recommendations in the 2000 *Dietary Guidelines for Americans*. However, in the 2005 *Dietary Guidelines*, recommendations on food consumption are expressed in cups (for fruits, vegetables, and dairy products) and ounce-equivalents (grains and meat) instead of servings. This does not affect the measurement of grain consumption, because one ounce-equivalent is identical to one serving for grain products. Therefore, the PSD is still directly applicable to the current recommendation on grain consumption.

GRAIN CONSUMPTION: WHAT KIND, BY WHOM, AND HOW MUCH?

The 2005 *Dietary Guidelines* included several changes in the recommendations for grain consumption. The most prominent are the quantitative recommendations for consumption of whole grains. Second, the recommendations for consumption of total grains were revised slightly downward. For example, recommended total grain consumption is now five 1-ounce equivalents (servings) per day for a 1,600-calorie diet, compared with six servings recommended in 2000. Third, the new guidelines cover a much wider range of food energy intakes, from 1,000 to 3,100 calories, compared with the 1,600–2,800 calories specified in the previous guidelines (figure 1).

Americans Favor Refined Grains over Whole Grains

The current recommendations for grain consumption are specified for 12 caloric levels (figure 1). We used linear interpolation to derive recommendations for intakes that fall between the specified levels. During 1994-96 and 1998, Americans age 2 and above consumed on average 1,987 calories per day, which corresponds to a recommendation of 6.3 ounces of total grains (table 1). During this period, Americans actually consumed 6.7 ounces of total grains per day, or 106 percent of the recommendation (figure 2). However, they overconsumed refined grains (5.6 ounces per day), averaging 77 percent over the recommended amount of 3.1 ounces, or half of total grains. It is a major challenge for Americans to meet the new guidelines for whole grains, as average consumption in the 1994-96 and 1998 surveys amounted to 1.1 ounces (34 percent of the recommended amount), and only 7 percent of consumers met the recommendation (table 1).

Table 1. U.S. grain consumption by gender and age

Intake and recommendation	All U.S	Females	Males	Children	All adults	Adults without children	Adults with children
Caloric intake (kcal/day)[1]	1987	1641	2349	1975	1991	1969	2020
Ounces/day							
Recommended intake of total grains	6.30	5.40	7.24	6.28	6.31	6.21	6.42
Whole grains[2]	3.15	2.70	3.62	3.14	3.15	3.11	3.21
Actual grain consumption:							
Total grains	6.68	5.62	7.79	6.73	6.66	6.64	6.64
Whole grains	1.07	0.94	1.21	1.02	1.09	1.11	1.00
At home	0.90	0.79	1.01	0.84	0.92	0.95	0.83
Away from home	0.17	0.15	0.19	0.17	0.17	0.16	0.17
Refined grains	5.61	4.69	6.58	5.72	5.57	5.52	5.64
At home	3.69	3.18	4.24	3.76	3.67	3.61	3.65
Away from home	1.92	1.51	2.34	1.96	1.90	1.92	1.99
Percent							
Consumption to recommendation:							
Total grains	105.54	104.21	106.94	106.53	105.17	106.45	102.87
Whole grains	34.30	34.98	33.59	32.43	35.01	36.97	30.62
Refined grains	176.79	173.44	180.29	180.63	175.34	175.92	175.11
Share of people meeting the recommendation:							
Total grains	53.45	52.94	53.99	54.80	52.95	53.74	50.93
Whole grains	7.34	7.32	7.35	5.41	8.06	8.72	6.68
Refined grains	87.42	86.75	88.12	91.28	85.97	85.59	86.89
Ounces/1,000 kcal							
Whole-grain density[3]:							
2005 Dietary Guidelines recommendation[2]	1.64	1.68	1.60	1.63	1.64	1.64	1.65
Reported whole-grain consumption	0.56	0.58	0.53	0.52	0.57	0.60	0.49
At-home consumption	0.68	0.71	0.65	0.64	0.70	0.75	0.59
Away-from-home consumption	0.29	0.31	0.27	0.29	0.29	0.28	0.30
Percent							
Away-from-home share of caloric intake	30.69	29.53	31.90	31.16	30.51	30.31	32.65

[1] Average of 2 days. [2] Half of total grains. [3] Whole-grain density is the recommended or reported consumption of whole grains per 1,000-calorie intake.

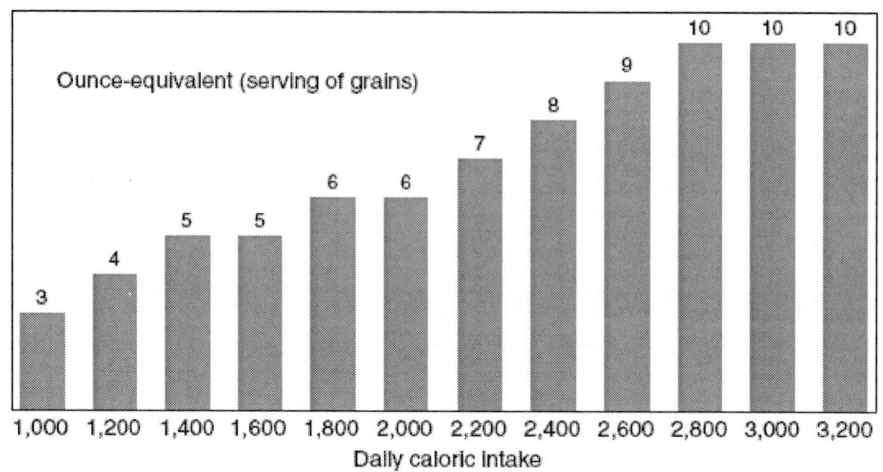

Source: Dietary Guidelines for Americans, 2005

Figure 1. Recommended consumption of grains: At least half the total should be whole grains

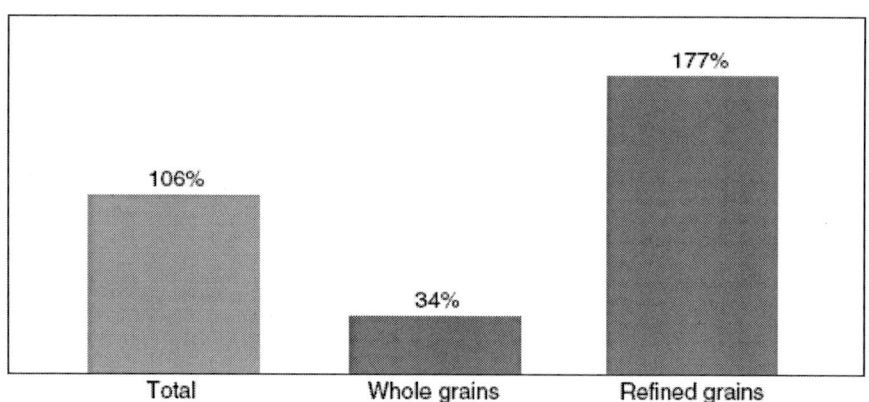

Source: CSFII 1994-96 and 1998.

Figure 2. U.S. grain consumption as a percent of 2005 recommendations

Recommendations for food consumption are based on caloric intakes, which vary by gender, age, physical activity, and body weight. On average, males (men and boys) consumed more whole grains than females (1.21 vs. 0.94 ounces), but females consumed a slightly higher percentage of the recommended level, 35 percent vs. 34 percent for males. Children age 2 to 19 consumed slightly more ounces of total grains than adults (6.73 vs. 6.66), but children appeared to favor refined grains over whole grains even more than adults did. Adults reached 35 percent of the recommended whole-grain intake, compared with 32 percent for children (figure 3). Further, the presence of children at home appeared to affect adults' grain consumption, especially of whole grains: compared with the recommended level, adults living with children consumed the same amount of refined grains as adults without children (176 and 175 percent of the recommended level), but adults living with children consumed less whole grain than other adults (31 percent of the recommendation vs. 37 percent).

Table 2. U.S. grain consumption by race and ethnicity

Intake and recommendation	Whites	Blacks	Hispanics	Asians	Others
Average caloric intake (kcal/day)[1]	2001	1963	1932	1969	1876
Ounces/day					
Recommended intake of total grains	6.36	6.07	6.20	6.27	6.08
Whole grains[2]	3.18	3.04	3.10	3.13	3.04
Actual grain consumption:					
Total grains	6.77	6.10	6.46	7.82	6.40
Whole grains	1.11	0.72	1.29	0.72	0.87
At home	0.94	0.59	1.09	0.62	0.70
Away from home	0.17	0.13	0.21	0.09	0.17
Refined grains	5.66	5.39	5.17	7.11	5.53
At home	3.73	3.35	3.48	5.05	3.65
Away from home	1.93	2.03	1.69	2.05	1.88
Percent					
Consumption to recommendation:					
Total grains	106.53	98.26	102.42	124.12	103.35
Whole grains	35.54	25.05	41.08	22.27	27.57
Refined grains	177.53	171.47	163.76	225.98	179.13
Share of people meeting the recommendation:					
Total grains	54.57	44.48	50.09	75.46	54.31
Whole grains	7.49	3.79	11.01	6.34	5.66
Refined grains	88.15	84.36	83.11	95.39	91.89
Ounces/1,000 kcal					
Whole-grain density[3]:					
2005 Dietary Guidelines recommendation[2]	1.63	1.65	1.65	1.65	1.69

Table 2. (Continued)

Intake and recommendation	Whites	Blacks	Hispanics	Asians	Others
Reported whole-grain consumption	0.57	0.41	0.66	0.36	0.45
At-home consumption	0.71	0.51	0.75	0.41	0.52
Away-from-home consumption	0.29	0.23	0.38	0.20	0.25
Percent					
Away-from-home share of caloric intake	30.90	31.56	29.19	27.87	28.95

1Average of 2 days. [2]Half of total grains. [3]Whole-grain density is the recommended or reported consumption of whole grains per 1,000-calorie intake.

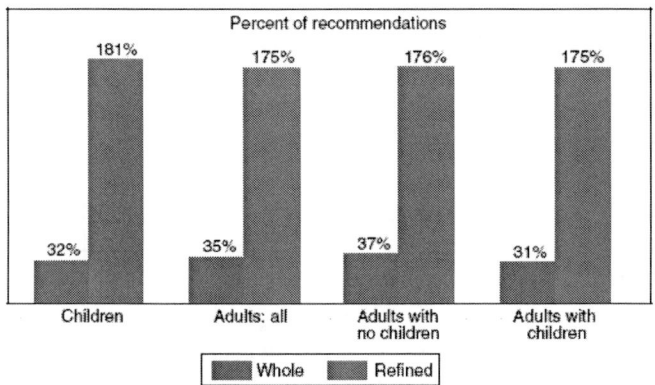

Source: CSFII 1994-96 and 1998.

Figure 3. U.S. grain consumption by children and adults

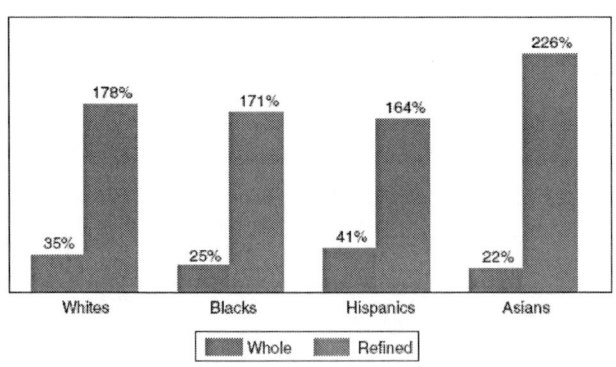

Source: CSFII 1994-96 and 1998. Table 2

Figure 4. U.S. grain consumption as a percent of recommendations, by race and ethnicity

Grain-type preference varies by race and ethnicity (table 2). Asians ate more grains than others, but they registered a strong preference for refined over whole grains, consuming more than double the recommended level of refined grains (226 percent) and having the lowest whole-grain intake among all consumers (figure 4). Hispanic consumers did better than other consumers in meeting the recommendation for whole-grain consumption, consuming 41 percent of the recommended level, compared with 35 percent for Whites, 25 percent for Blacks, and 22 percent for Asians. Eleven percent of Hispanics met the whole-grain recommendation, compared with 7 percent for Whites, 6 percent for Asians, and 4 percent for Blacks.

Table 3. U.S. grain consumption by income

Intake and recommendation	Income-poverty ratio			
	< 130%	131-185%	186-300%	> 300%
Average caloric intake (kcal/day)[1]	1937	1925	1946	2023
Ounces/day				
Recommended intake of total grains:	6.02	6.16	6.20	6.48
Whole grains[2]	3.01	3.08	3.10	3.24
Actual grain consumption:				
Total grains	6.44	6.31	6.43	6.96
Whole grains	0.95	0.98	0.98	1.17
At home	0.82	0.84	0.82	0.97
Away from home	0.13	0.15	0.15	0.20
Refined grains	5.49	5.33	5.45	5.79
At home	3.76	3.65	3.66	3.70
Away from home	1.73	1.68	1.79	2.09
Percent				
Consumption to recommendation:				
Total grains	104.53	102.37	103.42	107.57
Whole grains	31.27	32.18	32.20	36.84
Refined grains	177.80	172.55	174.63	178.30
Percent				
Share of people meeting the recommendation:				
Total grains	50.09	51.82	52.37	55.58
Whole grains	7.11	6.83	6.37	7.94
Refined grains	85.67	87.32	87.50	88.08

Table 3. (Continued)

Intake and recommendation	Income-poverty ratio			
	< 130%	131-185%	186-300%	> 300%
Ounces/1,000 kcal				
Whole-grain density[3]:				
2005 Dietary Guidelines recommendation[2]	1.66	1.64	1.64	1.63
Reported whole-grain consumption	0.51	0.52	0.52	0.59
At-home consumption	0.61	0.64	0.62	0.74
Away-from-home consumption	0.26	0.25	0.27	0.31
Percent				
Away-from-home share of caloric intake	26.86	28.32	29.92	33.03

[1] Average of 2 days. [2] Half of total grains. [3] Whole-grain density is the recommended or reported consumption of whole grains per 1,000-calorie intake.

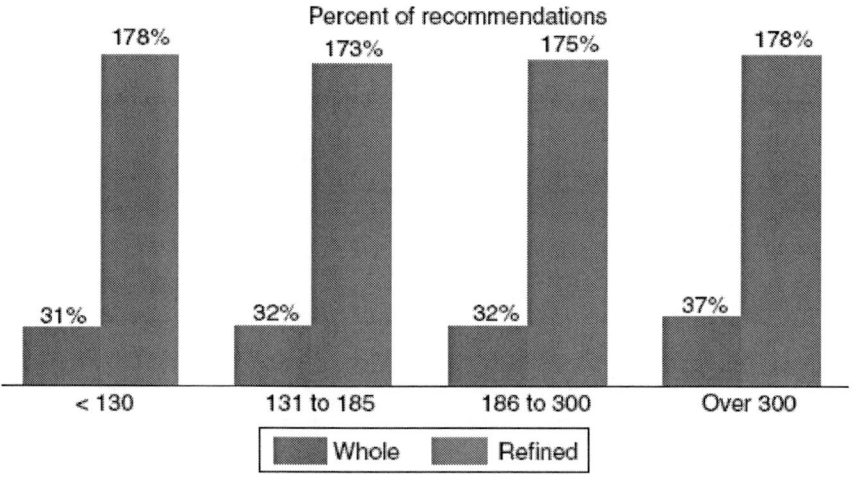

* as a ratio of income to poverty.
Source: CSFII 1994-96 and 1998.

Figure 5. U.S. grain consumption by income*

Grain consumption also varied by consumer income and education. In general, individuals with higher educational status earn higher incomes. (Household income is measured as a percentage of the Federal poverty guideline.) Consumers in the highest income bracket (300 percent of poverty or higher) ate 1.17 ounces of whole grains and 5.79 ounces of refined grains, reaching 37 percent and 178 percent of the recommended whole-grain and refined-grain servings, compared with 31 and 178 percent for individuals in the lowest income bracket (130 percent of poverty or lower, see table 3). College-educated individuals consumed 38 percent of the whole-grain recommendation, compared with 32 percent for those without a high school diploma, 30 percent for high school graduates, and 35 percent for those who attended college without completing a degree (table 4).

Table 4. U.S. grain consumption by educational level

Intake and recommendation	< High school	High school	Some college	College
Average caloric intake (kcal/day)[1]	1996	1981	2072	1949
Ounces/day				
Recommended intake of total grains	6.16	6.31	6.47	6.27
Whole grains[2]	3.08	3.15	3.24	3.14
Actual grain consumption:				
Total grains	6.68	6.46	6.82	6.76
Whole grains	0.97	0.93	1.12	1.17
At home	0.85	0.79	0.91	0.98
Away from home	0.12	0.14	0.21	0.19
Refined grains	5.71	5.53	5.70	5.59
At home	3.81	3.68	3.56	3.72
Away from home	1.90	1.85	2.14	1.87
Percent				
Consumption to recommendation:				
Total grains	106.41	102.40	104.99	107.47
Whole grains	31.90	30.08	35.42	37.51
Refined grains	180.91	174.73	174.56	177.43
Share of people meeting the recommendation:				
Total grains	106.41	102.40	104.99	107.47
Whole grains	31.90	30.08	35.42	37.51
Refined grains	180.91	174.73	174.56	177.43
Ounces/1,000 kcal				
Whole-grain density[3]:				
2005 Dietary Guidelines recommendation[2]	1.65	1.65	1.62	1.64

Table 4. (Continued)

Intake and recommendation	< High school	High school	Some college	College
Reported whole-grain consumption	0.52	0.49	0.57	0.61
At-home consumption	0.64	0.60	0.69	0.75
Away-from-home consumption	0.22	0.26	0.30	0.33
Percent				
Away-from-home share of caloric intake	27.32	30.47	34.24	30.63

[1] Average of 2 days. [2] Half of total grains. [3] Whole-grain density is the recommended or reported consumption of whole grains per 1,000-calorie intake.

Detecting the Source of Whole-Grain Deficiency by Density Measurement

With an average intake of 1,987 calories a day during 1994-96 and 1998, Americans needed to consume 3.15 ounces of whole grains in order to meet the 2005 recommendation. Therefore, each 1,000 calories of energy intake needed to incorporate 1.64 ounces of whole grains. Whole-grain density, measured as ounces of whole grains consumed per 1,000-calorie intake, is a useful yardstick for comparing whole-grain consumption across population subgroups, as well as for detecting sources of dietary deficiency.

During 1994-96 and 1998, Americans consumed only 0.56 ounce of whole grains per 1,000 calories, slightly above one-third of the recommendation (table 1). Judging from the whole-grain density, females did better than males (0.58 vs. 0.53 ounce), adults did better than children (0.57 vs. 0.52 ounce), and Asians and Blacks were behind other consumers (0.36 and 0.41 ounce vs. 0.57 ounce for Whites and 0.66 ounce for Hispanics) in reaching whole-grain recommendations (table 2). Whole-grain density rose with educational achievement and household income (tables 3 and 4), consistent with the consumption-to-recommendation ratios.

Eating Out Poses a Challenge to Whole-Grain Consumption

The most significant lifestyle change of the past two decades in the United States is probably the increase in dining out. Americans consumed about a third of their calories from food prepared away from home during 1994-96 and 1998, up from less than a fifth in 1977-78 (USDA/ERS, 2007a). But when people order items from menus, whole grains seldom make the list. The whole-grain density for food prepared away from home is low compared with that for food prepared at home (0.29 vs. 0.68 ounce per 1,000 calories; see table 1). Therefore, the rising popularity of eating out could present a barrier to incorporating more whole grains into our diets.

Females did better than males in including whole grains in their diets, both at home and away from home. There were few differences in the whole-grain density in the away-from-home foods consumed by children and adults, but greater variation in foods prepared at home. Compared with all other consumers, Hispanics did much better in incorporating whole grains in their meals when they ate out (with a density of 0.38 ounce per 1,000 calories, vs. 0.29 ounce for Whites, 0.23 ounce for Blacks, and 0.2 ounce for Asians) (figure 6 and table 2). College-educated adults did better than other adults in incorporating whole grains into meals eaten both at home and away from home (table 4).

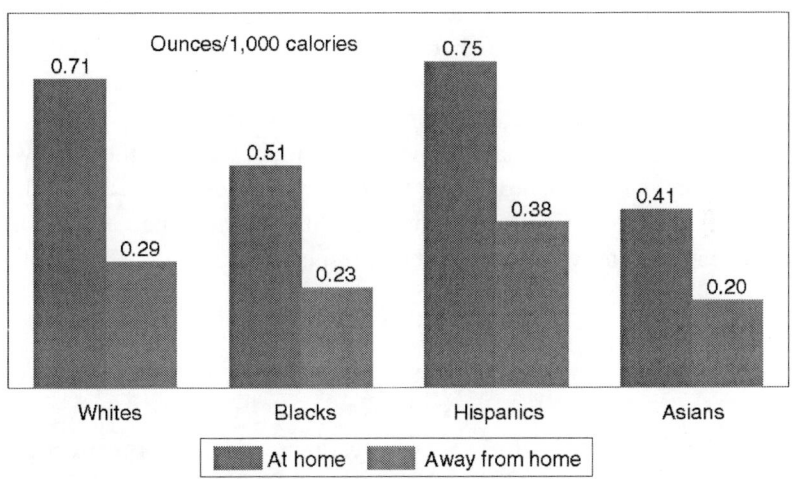

Source: CSFII 1994-96 and 1998.

Figure 6. Whole-grain content in U.S. meals prepared at home and away, by race and ethnicity

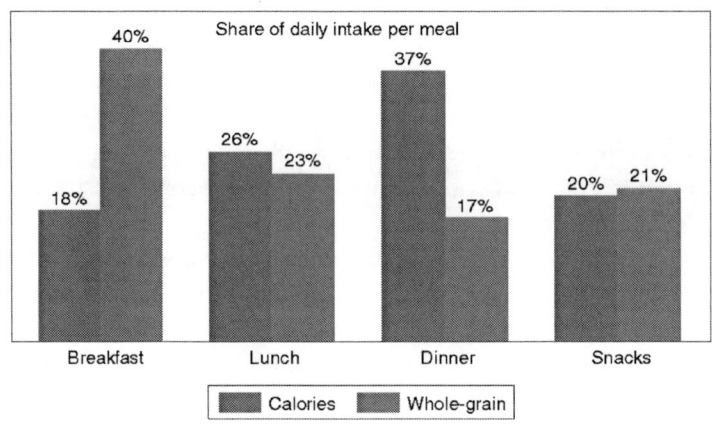

Source: CSFII 1994-96 and 1998.

Figure 7. Breakfast provided more whole grains than other meal occasions

CSFII respondents reported the meal occasion and time for each food eaten. Meal time was classified into four categories: breakfast (breakfast and brunch eaten before 10 a.m.), lunch (brunch after 10 a.m. and lunch), dinner, and snacks. Breakfast accounted for 18 percent of Americans' daily caloric intake, but contributed 40 percent of whole-grain consumption (figure 7). Lunch contributed 26 percent of caloric intake and 23 percent of whole- grain consumption. On the other hand, dinner—the main meal of the day, with 37 percent of daily energy intake during the survey period— contributed the least (17 percent) to whole-grain consumption. Snacks accounted for 20 and 21 percent of caloric and whole-grain intake, respectively.

AN ECONOMETRIC MODEL OF GRAIN CONSUMPTION

An objective of this study was to estimate the factors affecting U.S. consumption of refined and whole grains. Due to the nature of the data and the model specification, we estimated a censored demand system with endogenous regressors. The econometric model, which is quite technical, is included in the appendix.

We specified a four-equation demand system to estimate factors affecting the consumption of refined and whole grains. Even though grains are staple foods in American diets, the U.S. consumption of whole grains is low, and many consumers do not eat whole grains on a given day. CSFII respondents reported their dietary intakes for 2 nonconsecutive days. During 1994-96 and 1998, among those age 2 and above, 24 percent did not eat any whole grains over the 2 dietary recall days. This is the source of the censored dependent variable in demand estimation.

The food demand literature has shown that food and nutrient intakes are affected by dietary knowledge. Since the passage of the 1990 Nutritional Labeling and Education Act (NLEA) and the release of the 1994–96 DHKS data, researchers have paid increasing attention to the effects of dietary knowledge, attitude, and food-label use on the intake of foods and nutrients and on diet quality. The NLEA mandates that the Nutrition Fact Panel be affixed to packaged foods. Use of the panel has been found to affect the intake of fat (Kreuter and Brennan, 1997; Neuhouser et al., 1999) and cholesterol, sodium, and fiber (Kim et al., 2000).

Dietary knowledge and attitudes have been linked to consumption of fats and oils (Kim and Chern, 1995), fat-modified foods (Coleman and Wilson, 1994), eggs (Brown and Schrader, 1990; Kan and Yen, 2003; Yen et al., 1996), meat (Kaabia et al., 2001; Kinnucan et al., 1997), and 25 food groups consumed at and away from home (Lin et al., 2003). There are reported links between knowledge and intake of fiber (Variyam et al., 1996), energy and nutrient density (Bhargava, 2004), and fat and cholesterol (Variyam et al., 1997, 1999a). Dietary knowledge and attitude have also been linked to the diet quality of children (Variyam et al., 1999b), the elderly (Howard et al., 1998), and women household heads (Ramezani and Roeder, 1995).

In the last *Nationwide Food Consumption Survey*, conducted in 1987-88, USDA collected data on both household food use and individual food intake. In the household food use component of the survey, data on both quantity and spending were collected for foods purchased by each household. Since 1987/88, the household food use questions have been discontinued, and only data on individual food intake have been collected. Consequently, only food intake—but not spending, and hence not unit value—is available in recent food consumption surveys. Our demand specification thus suffers from missing price variables, even though regional and seasonal variables are included to capture some systematic price variations across regions and seasons.

The belief that white flour was the food of the rich and unrefined flour the food of hard-working peasants and the poor is a plausible explanation for the low consumption of whole grains relative to refined grains in the United States. Low consumption could also be attributed to lack of availability, tastes and preferences, and/or higher prices. Historically, some whole-grain products have been more expensive because they were specialty items produced in smaller quantities (Buzby et al., 2005). Therefore, the higher cost of manufacturing and marketing whole grains relative to the cost of refined grains, which benefit from economies of scale, further dampens the demand for whole grains.

An analysis of supermarket scanner data by Frazao and Allshouse (1996) found that the average price of whole-grain products was higher than the price of refined-grain products ($1.80 vs. $1.36 per pound) in 1995. Whole-grain bread and brown rice were sold at $1.19 and $1.13 per pound, compared with $0.99 and $0.68 for refined-grain bread and white rice. Similar price differentials have been reported more recently (Buzby et al., 2005; Kantor et al., 2001). If the price differential is indeed due to economies of scale, then an expanding whole-grain market—sparked by greater demand—would ease existing price differentials and further increase the demand for whole grains.

In this study, we hypothesize that consumer knowledge and attitudes and food-label use affect the consumption of refined and whole grains. Unlike sociodemographic factors, consumer knowledge, attitudes, and behavior are likely to be determined by the same factors that determine consumption. We accommodate this data feature by treating knowledge and attitudes as endogenous choice variables in a censored demand system. Such an econometric specification is new to the applied literature.

Data and Model Estimation

CSFII collected socioeconomic and demographic data for the sample households and their members. The socioeconomic and demographic variables that are hypothesized to affect grain consumption in this study include household income, household size, household structure, gender, age, race/ethnicity, household region, and season (table 5).

We hypothesize that the use of nutrition labels and the perceived importance of consuming plenty of grain products also affect grain consumption, and these two variables are also treated as endogenous in the censored demand system. In addition to income, gender, age, and race/ethnicity, the use of nutrition labels and perceived importance of grain consumption are hypothesized to be affected by education, exercise, smoking status, whether the respondent is a meal planner, and whether anyone in the household is on a special diet.

Table 5. Variable definitions and sample statistics (n = 5,501)

Variable	Definition	Mean
Endogenous variables:		
Label use	Use the short phrases on the label like 'low fat' or 'light' or 'good source of fiber': 1 = often or sometimes; 0 = rarely or never.	0.61
Importance	Perceived importance in choosing a diet with plenty of breads, cereals, rice, and pasta: 1 = very or somewhat; 0 = not too important or not at all	0.74
Whole grains	Daily consumption of whole grains (servings), 2-day average	0.28 (0.34)
	Consuming sample (n = 4003)	0.38 (0.34)
Refined grains	Daily consumption of refined grains (servings), 2-day average	1.34 (0.61)
Household (HH) characteristics: continuous exogenous variables:		
Income	Household income as percent of poverty	160.90 (137.10)
HH size	Number of persons in the household	2.56 (1.46)
Binary exogenous variables (yes = 1; no = 0)		
Household characteristics:		
HH type 1	Household is dual-headed, with children	0.28
HH type 2	Household is dual-headed, without children	0.36
HH type 3	Household is single-headed, with children	0.08
HH type 4	Household is single-headed without children (reference)	0.28
Special diet	A family member is on a special diet	0.27
Individual characteristics:		
Male	Respondent is male	0.50
Age 20–30		0.14
Age 31–40		0.18
Age 41–50		0.18

Table 5. (Continued)

Variable	Definition	Mean
Age 51–60		0.18
Age > 60	Age 61 or older (reference)	0.32
Black	A non-Hispanic Black	0.11
Hispanic	Of Hispanic origin	0.08
Asian	Asian/Pacific Islander	0.02
Other race	None of the above nor White	0.01
White	Non-Hispanic White (reference)	0.78
Quarter 1	Dietary recalls taken in January-March	0.23
Quarter 2	Dietary recalls taken in April-June	0.26
Quarter 3	Dietary recalls taken in July-September	0.28
Quarter 4	Dietary recalls taken in October-December (reference)	0.24
Midwest	Resides in a Midwestern State	0.25
South	Resides in a Southern State	0.35
West	Resides in a Western State	0.20
Northeast	Resides in a Northeastern State (reference)	0.19
Rural	Resides in a rural area	0.27
Suburb	Resides in a suburb	0.44
City	Resides in the central city (reference)	0.30
< high school	Did not complete high school (reference)	0.22
High school	Completed high school education	0.34
Some college	Attended college for less than 4 years	0.21
College	Had 4 or more years of college education	0.23
Meal planner	Main meal planner of household	0.70
Exercise	Exercised vigorously: at least twice a week	0.48
Smoker	Currently smoking cigarettes	0.26
Pessimistic	Agrees with statement that some are born to be fat and some thin	0.43

Note: Standard deviations in parentheses.

In the *1994-96 Diet and Health Knowledge Survey* (DHKS), respondents were asked about the perceived importance of choosing a diet with plenty of breads, cereals, rice, and pasta. The answers were grouped into important (very or somewhat) and not important (not too or not at all). The respondents were also asked if, when they buy foods, whether they often, sometimes, rarely, or never use the information in (1) the short phrases on the label like "low fat" or "light" or "good source of fiber," (2) the list of ingredients, (3) the nutrition panel listing the amount of nutrients, and (4) claims for health benefits of nutrients or foods. These four possible answers are grouped into "use" (often or sometimes) and "do not use" (rarely or never). The four types of label use are examined in the model estimation.

Two of the endogenous variables (perception and label use) and many exogenous variables come from the DHKS, which surveyed only adults; hence, our analysis is limited to the adult sample in the 1994-96 CSFII. Excluding observations with missing values, the final sample for the regression analysis contains 5,501 adults. Of the sample, 72.8 percent consumed whole-grain products, while almost all individuals (99.8 percent) consumed refined-grain products. Sample means are reported in table 5, along with the variable definitions.

Empirical Results

The four-equation system, consisting of binary equations for food label use and perceived importance of grains and censored equations for whole and refined grains, was estimated by maximizing the likelihood function described in the appendix. Four alternative variables represent food labels— the list of ingredients, short phrases, nutrition panel, and health claims. These alternative specifications of label use produce similar results. For brevity, we present only the results from reading short phrases. This is because a short-phrase example used in the interview is related to the fiber content of foods. Whole grains are known for their rich fiber content.

The empirical results suggest that use of food labels and consumer attitude should both be treated as endogenous, and that the two binary equations and two consumption equations should be estimated as a system. Maximum-likelihood estimates for the system are reported in table 6. At the 10- percent-or-better level of significance, two-thirds of the variables are significant in the label-use and whole-grain equations, and about half are significant in the perceived importance and refined-grain equations.

Table 6. Parameter estimates of demands for whole and refined grains with endogenous food label use and perceived importance

Variable	Label use		Importance of grains		Whole grains		Refined grains	
	Coeff.	S.E.	Coeff.	S.E.	Coeff.	S.E.	Coeff.	S.E.
Constant	0.186***	0.070	0.484***	0.070	−0.208***	0.054	1.190***	0.070
Income	0.487***	0.146	−0.211	0.155	0.094	0.059	0.060	0.074
Male	−0.481***	0.043	0.010	0.043	0.042***	0.017	0.148***	0.024
Age 20–30	−0.061	0.061	0.149***	0.060	−0.061***	0.022	0.156***	0.028
Age 31–40	0.023	0.056	0.122**	0.057	−0.054***	0.022	0.124***	0.028
Age 41–50	0.074	0.055	0.123**	0.055	−0.065***	0.020	0.053**	0.027
Age 51–60	0.114**	0.054	0.069	0.054	−0.055***	0.019	0.046*	0.025
Black	0.007	0.059	−0.208***	0.056	−0.118***	0.022	−0.091***	0.029
Asian	0.171	0.155	−0.124	0.149	−0.253***	0.054	0.487***	0.054
Other race	0.251	0.181	0.269	0.196	−0.117**	0.060	−0.037	0.077
Hispanic	0.117*	0.068	−0.070	0.068	0.028	0.023	0.048	0.032
Special diet	0.224***	0.041	0.051	0.042	0.021	0.016	−0.018	0.020
High school	0.168***	0.051	0.067	0.047				
Some college	0.190***	0.058	0.125**	0.055				
College	0.323***	0.061	0.368***	0.060				
Meal planner	0.075*	0.045	0.133***	0.043				
Exercise	0.126***	0.037	0.134***	0.035				
Smoker	−0.273***	0.042	−0.163***	0.040				
Pessimistic	−0.153***	0.037	−0.058*	0.035				
Midwest					0.043**	0.019	−0.041*	0.025
South					−0.005	0.018	−0.087***	0.023
West					0.137***	0.020	−0.191***	0.027
Rural					−0.034**	0.016	−0.077***	0.023
Suburban					−0.025*	0.015	−0.010	0.020
HH type 1					−0.055**	0.026	0.008	0.035
HH type 2					−0.027*	0.017	0.002	0.022
HH type 3					−0.095***	0.028	−0.031	0.042
Quarter 1					0.015	0.017	0.031	0.023
Quarter 2					0.003	0.017	−0.027	0.023
Quarter 3					−0.023	0.016	−0.044**	0.022
HH size					0.078	0.063	0.050	0.088
Label use					0.168***	0.061	−0.059	0.088
Importance					0.421***	0.059	0.210**	0.094
Std. dev. (i)					0.451***	0.011	0.592***	0.008
Error correlation								
Importance	0.163***	0.025						
Whole grains	−0.169**	0.083	−0.510***	0.065				
Refined grains	0.068	0.088	−0.156*	0.087	−0.178***	0.023		

Note: Log-likelihood value = −14611.41 2. Asterisks ***, **, and * indicate significance at 1%, 5%, and 10% levels, respectively. HH = household.

Asians have a stronger preference for refined grains over whole grains than do other ethnic groups. In fact, marginal effects (table 7) suggest that being of Asian descent is one of the most influential factors in grain consumption.

Relative to Whites, Asians are 22 percent less likely to consume whole grains. Those Asians who do eat whole grains consume 0.1 fewer servings per day than White whole-grain eaters. Compared with Whites, Blacks are less likely to perceive grain consumption as important, and they tend to consume less of both refined and whole grains (table 6). There are few differences between Whites and Hispanics in terms of refined- and whole- grain consumption. Adults who live with a family member with special dietary needs are more inclined to use food labels, and hence to consume more whole grains.

Regional differences are evident, with Midwestern and Western residents consuming more whole grains than individuals in other parts of the United States, and individuals in the Midwest, South, and West consuming fewer refined grains than those in the Northeast. Urbanization also plays a role, with rural residents consuming less of both whole and refined grains, and suburban residents consuming fewer whole grains, than individuals in cities.

Household structure is classified into four categories: dual- or single-headed and with or without children, with single-person households the reference group. Respondents from households with children (dual- or single-headed) eat fewer servings of whole grains. This is consistent with past findings that children prefer white bread (Harnack et al., 2003; Moutou et al., 1998).

CONCLUSION: INCREASING WHOLE-GRAIN CONSUMPTION IS AN UPHILL BATTLE

Despite longstanding Government efforts to promote the consumption of whole grains, the American diet is still far short of the goals. Americans consume about 2,000 calories a day and need to eat 6 servings of grains, with at least 3 as whole grains. This translates into at least 1.5 servings of whole grains per 1,000 calories. During the study period, Americans ate only 0.56 serving of whole grains in that quantity of calories.

The American fondness for eating out, a fast-growing trend, may be an obstacle. Whole grains are particularly lacking in meals eaten in food establishments. Over the study period, there was only 0.29 serving of whole grains in each 1,000 calories consumed outside the home. With the current whole-grain content in away-from-home foods, Americans need to eat over 10,000 calories per outside meal to get the recommended 3 servings of whole grains prescribed for each 1,000 calories. Thus, unless appealing whole-grain dishes become more available in restaurants, the popularity of eating out could further erode the whole-grain base in American diets.

REFERENCES

Amemiya, T. (1974). "Multivariate Regression and Simultaneous Equation Models When the Dependent Variables are Truncated Normal," *Econometrica, 42*, 999-10 12.

Bhargava, A. (2004). "Socio-Economic and Behavior Factors are Predictors of Food Use in the National Food Stamp Program Survey," *British Journal of Nutrition, 92*, 497-506.

Brown, D.J. & Schrader, L. F. (1990). "Cholesterol Information and Shell Egg Consumption," *American Journal of Agricultural Economics 72*, 548-55.

Buzby, J. C., Farah, H. A. & Vocke, G. (2005). "Will 2005 Be the Year of the Whole Grain?" *Amber Waves,* U.S. Department of Agriculture, *Economic Research Service, 3(3)*, 12-17.

Catsiapis, G. & Robinson, C. (1982). "Sample Selection Bias with Multiple Selection Rules: An Application to Student Aid Grants," *Journal of Econometrics, 18*, 351-68.

Centers for Disease Control and Prevention (CDC), National Center for Health Statistics. *National Health and Nutrition Examination Survey* (NHANES). U.S. Department of Health and Human Services. *http://www.cdc.gov/nchs/nhanes.htm.*

Coleman, L. M. & Wilson, M. A. (1994). "Consumers' Knowledge and Use of Fat-Modified Products," *Journal of Family and Consumer Sciences, 4*, 26-33.

Frazao, E. & Allshouse, J. (1996). *Size and Growth of the Nutritionally Improved Foods Market.* U.S. Department of Agriculture, Economic Research Service, Food and Consumer Economics Division, Agriculture Information Report No. 723, April.

Harnack, L., Walters, S. A. & Jacobs, D. R. Jr. (2003). " Dietary Intake and Food Sources of Whole Grains Among U.S. Children and Adolescents:

Data from the 1994-96 Continuing Survey of Food Intakes by Individuals," *Journal of the American Dietetic Association, 103*, 1015-19.

Howard, J. H., Gates, G. E., Ellersieck, M. R. & Dowdy, R. P. (1998). "Investigating Relationships Between Nutritional Knowledge, Attitudes, and Beliefs, and Dietary Adequacy of the Elderly," *Journal of Nutrition for the Elderly, 17*, 35-54.

Kaabia, M. B., Angulo, A. M. & Gil, J. M. (2001). "Health Information and the Demand for Meat in Spain," *European Review of Agricultural Economics, 28*, 499-517.

Kan, K. & Yen, S. T. (2003). "A Sample Selection Model with Endogenous Health Knowledge: Egg Consumption in the USA," in *Health, Nutrition and Food Demand*, K. Rickertsen and W.S. Chern (eds). CABI Publishing, Cambridge, MA, 91-103.

Kantor, L. S., Variyam, J. N., Allshouse, J. E., Putnam, J. J. & Lin, B. (2001). "Choose a Variety of Grains Daily, Especially Whole Grains: a Challenge to Consumers," *Journal of Nutrition, 131*, 473S-486S.

Keane, M. & Moffitt, R. (1998). "A Structural Model of Multiple Welfare Program Participation and Labor Supply," *International Economic Review, 39*, 553-89.

Kim, S., Nayga, R. M. Jr. & Capps, O. Jr (2000). "The Effect of Food Label Use on Nutrient Intakes: An Endogenous Switching Regression Analysis," *Journal of Agricultural and Resource Economics, 25*, 215-31.

Kim, S. R. & Chern, W. S. (1999). "Alternative Measures of Health Information and Demand for Fats and Oils in Japan," *Journal of Consumer Affairs, 33*, 92-109.

Kinnucan, H. W., Xiao, H., Hsia, C. J. & Jackson, J. D. (1997). "Effects of Health Information and Generic Advertising on U.S. Meat Demand," *American Journal of Agricultural Economics, 79*, 13-23.

Kotz, S., Johnson, N. & Balakrishnan, N. (2000). *Continuous Multivariate Distributions*, vol. 1: *Models and Applications*, 2nd ed., John Wiley & Sons, New York.

Kreuter, M. W. & Brennan, L. K. (1997). "Do Nutrition Label Readers Eat Healthier Diets: Behavioral Correlates of Adults' Use of Food Labels," *American Journal of Preventive Medicine, 13*, 277-83.

Liebman, B. (1997). "The Whole Grain Guide," *Nutrition Action Health Letter*. March – U.S. Edition. Washington, DC: Center for Science in the Public Interest. Available at: http://www.cspinet.org/nah/wwheat.html. Accessed 18 May 2005.

Lin, B., Variyam, J. N., Allshouse, J. & Cromartie, J. (2003). *Food and Agri-*

cultural Commodity Consumption in the United States: Looking Ahead to 2020. U.S. Department of Agriculture, Economic Research Service, Food and Rural Economics Division, Agricultural Economic Report No. 820, February.

Marquart, L., Slavin, J. & Fulcher, G. (eds.) (2002). *Whole-Grain Foods: In Health and Disease.* American Association of Cereal Chemists, Inc., St. Paul, MN.

Moutou, C., Brewster, G. W. & Fox, J. A. (1998). "U.S. Consumers' Socioeconomic Characteristics and Consumption of Grain-Based Foods," *Agribusiness, 14*, 63-72.

Neuhouser, M. L., Kristal, A. R. & Patterson, R. E. (1999). "Use of Food Nutrition Labels Is Associated With Lower Fat Intake," *American Journal of the Dietetic Association, 99*, 45-53.

Ramezani, C. A. & Roeder, C. (1995). "Health Knowledge and Nutrition Adequacy of Female Heads of Households in the United States," *Journal of Consumer Affairs, 29*, 381-402.

Spiller, G. A. (2002). "Whole Grains, Whole Wheat, and White Flours in History." Chapter 1 in *Whole-Grain Foods in Health and Disease.* L. Marquart, J.L. Slavin, and G. Fulcher (eds.). American Association of Cereal Chemists, Inc., St. Paul, MN.

Tunali, I. (1986). "A General Structure for Models of Double-Selection and an Application to a Joint Migration-Earnings Process with Remigration," *Research in Labor Economics, 8*, 235-84.

U.S. Department of Agriculture (2005). MyPyramid.gov: *Steps to a Healthier You.* Available at: http://www.mypyramid.gov/. Accessed 6 May 2005.

U.S. Department of Agriculture, Agricultural Research Service (2000). *Continuing Survey of Food Intakes by Individuals 1994–96 and 1998.* CD-ROM

U.S. Department of Agriculture, Economic Research Service. *Diet and Health Food Consumption and Nutrient Intake Tables.* Available at *http://www.ers.usda.gov/briefing/dietand* health/data/. Accessed 22 February 2007a.

U.S. Department of Agriculture, Economic Research Service. *Flour and Cereal Products: Per Capita Consumption.* Available at *http://www.ers.usda.gov/Data/FoodConsumption/FoodAvailIndex.htm.* Accessed 22 February 2007b.

U.S. Department of Agriculture and U.S. Department of Health and Human Services (USDHHS) (1990, 1995, 2000). *Nutrition and Your Health: Dietary Guidelines for Americans.* Home and Garden Bulletin No 232.

U.S. Department of Agriculture and U.S. Department of Health and Human Services (USDHHS) (2005). *Nutrition and Your Health: Dietary Guidelines for Americans, 2005.* Available at: *http://www.healthierus.gov/dietaryguidelines/*. Accessed 6 May 2005.

U.S. Department of Health and Human Services (USDHHS) (2005). *Healthy People 2010.* Available at: http://healthypeople.gov/. Accessed 6 May 2005.

U.S. Department of Health and Human Services, Food and Drug Administration (2007). *How To Understand and Use Nutrition Facts Labels.* Available at: http://www.cfsan.fda.gov/~dms/foodlab.html. Accessed 12 April 2007.

Variyam, J. N., Blaylock, J. R. & Smallwood, D. M. (1996). "Modeling Nutrition Knowledge, Attitudes, and Diet-Disease Awareness: The Case of Dietary Fiber," *Statistics in Medicine, 15*, 23-35.

Variyam, J. N., Blaylock, J. R. & Smallwood, D. M. (1997). *Diet-Health Knowledge and Nutrition: The Intake of Dietary Fats and Cholesterol.* U.S. Department of Agriculture, Economic Research Service, Technical Bulletin No. 1855.

Variyam, J. N., Blaylock, J. R. & Smallwood, D. (1999a). "Information, Endogeneity, and Consumer Health Behavior: Application to Dietary Intakes," *Applied Economics, 31*, 217-26.

Variyam, J. N., Blaylock, J. R., Lin, B., Ralston, K. & Smallwood, D. (1999b), "Mother's Nutrition Knowledge and Children's Dietary Intakes," *American Journal of Agricultural Economics, 8*, 373-84.

Wiemer, K. L. (2002). "Whole-Grains Health Claims: Supporting Scientific Evidence and the FDA Modernization Act Process." In *Whole-Grain Foods in Health and Disease*, L. Marquart, J.L. Slavin, and R.G. Fulcher (eds.). American Association of Cereal Chemists, Inc., St. Paul, MN.

Yen, S. T., Jensen, H. & Wan, Q. (1996). "Cholesterol Information and Egg Consumption in the U.S.: A Nonnormal and Heteroscedastic Double-Hurdle Model," *European Review of Agricultural Economics, 23*, 343-56.

APPENDIX: ECONOMETRIC MODEL OF GRAIN CONSUMPTION

Econometric Model

We developed an estimation procedure for an econometric system with censored dependent variables and endogenous regressors. In the equations that follow, observation subscripts are suppressed for brevity. The endogenous regressors, food label use (y_1) and nutrition knowledge (y_2), are both binary, and therefore specified as probit:

$$y_i = 1(z'\alpha_i + u_i > 0), \; i = 1,2 \qquad (1)$$

The censored equations for whole grains (y_3) and refined grains (y_4) are

$$y_i = \max(0, x'\beta_i + \gamma_{i1} y_1 + \gamma_{i2} y_2 + u_i), \; i = 3,4 \qquad (2)$$

In the above, $1(.)$ is a binary indicator function, z and x are exogenous vectors of explanatory variables, α_i and β_i are conformable vectors of parameters, γ_{i1} and γ_{i2} are scalar parameters, and the error terms $e \equiv [u_1, u_2, u_3, u_4]'$ are distributed as $e \sim \mathcal{N}(0, \Sigma)$ with probability density function (pdf) $f(u_1, u_2, u_3, u_4)$. The covariance matrix Σ is defined with error correlations ρ_{ij} and standard deviations σ_i so that $\sigma_1^2 = \sigma_2^2 = 1$; these parametric restrictions on σ_1 and σ_2 are necessary because variables y_1 and y_2 are both binary.

The model considered here is an extension of the Tobit system of Amemiya (1974), in that endogenous variables are present in Equation (2). It also generalizes the multiple- and double- selection models (Catsiapis and Robinson, 1982; Tunali, 1986) and multiple-treatment models (Keane and Moffitt, 1998) in that there are multiple outcomes in Equation (2), and that, in addition, these outcome variables are censored.

The Log-Likelihood Function

In this study, three different likelihood functions were constructed for different 'sample regimes,' characterized by zero/positive outcomes of whole grains and refined grains, as described below

When Both Whole Grains and Refined Grains Are Censored

This is the least observed case in which an individual does not consume either refined or whole grains. To facilitate description of the estimation procedure, define dichotomous indicators

$\kappa_1 = 2y_1 - 1$, $\kappa_2 = 2y_2 - 1$, a diagonal matrix $D = \text{diag}(\kappa_1, \kappa_2, -1, -1)$ and vector $r = [r_1, r_2, r_3, r_4]$,

where $r_i = z'\alpha_i$ for $i = 1, 2$ and $r_i = x'\beta_i + \gamma_{i1}y_1 + \gamma_{i2}y_2$ for $i = 3, 4$. The sample regime is

characterized by inequalities $e^* \leq Dr$, where $e^* = -De = [u_1^*, u_2^*, u_3, u_4]' \sim \mathcal{N}(0, \Omega)$ so that

$\Omega = D\Sigma D'$, for which the likelihood contribution is

$$L_1 = \int_{-\infty}^{\kappa_1 r_1} \int_{-\infty}^{\kappa_2 r_2} \int_{-\infty}^{-r_3} \int_{-\infty}^{-r_4} f(u_1^*, u_2^*, u_3, u_4) \, du_4 du_3 du_2^* du_1^*$$

$$= \int_{e^* \leq Dr} f(e^*) de^* \tag{3}$$

When Whole Grains Are Censored

This is the most frequently observed case, in which an individual consumed refined grains but not whole grains. Using Equations (1) and (2) and following a procedure similar to developing Equation (3), the likelihood contribution can be expressed as

$$L_2 = g(u_4) \int_{-\infty}^{\kappa_1 r_1} \int_{-\infty}^{\kappa_2 r_2} \int_{-\infty}^{-r_3} h(u_1^*, u_2^*, u_3 \mid u_4) \, du_3 du_2^* du_1^* \tag{4}$$

Where $g(u_4)$ is the marginal density of u_4, $h(u_1^*, u_2^*, u_3 \mid u_4)$ is the conditional density, and $u_4 = y_4 - (x'\beta_4 + \gamma_{71}y_1 + \gamma_{72}y_2)$. The likelihood contribution for a regime with refined grains censored follows from Equation (4) with equation subscripts 3 and 4 reversed.

When Whole Grains and Refined Grains Are Both Consumed

The likelihood contribution is

$$L_3 = g(u_3, u_4) \int_{-\infty}^{\kappa_1 r_1} \int_{-\infty}^{\kappa_2 r_2} h(u_1^*, u_2^* \mid u_3, u_4) \, du_2^* du_1^* \tag{5}$$

where $u_i = y_i - (x'\beta_i + \gamma_{i1}y_1 + \gamma_{i2}y_2), i = 3, 4$, and moments of the marginal distribution $g(u_3, u_4)$ and conditional distribution $h(u_1^*, u_2^* | u_3, u_4)$ are similar to those in Equation (4), with partition of e^* and its covariance matrix Ω at the second row.

The sample likelihood function for the system is the product of the likelihood contributions L_1, L_2 or L_3 over the sample, depending on the regime for each observation. The integrals in Equations (3), (4), and (5) can be evaluated by simulation or quadrature. The model reduces to an exogenous model when error correlations are zero between the two binary equations (foodlabel use and nutrition knowledge) and those of the consumption equations. The corresponding parametric restrictions are

$$\rho_{i,j} = 0 \quad \forall \quad i = 3, 4; j = 1, 2 \tag{6}$$

and a test for these parametric restrictions amounts to one for the joint endogeneity of y_1 and y_2. The restricted model implied by Equation (6) can be estimated by a bivariate probit for Equation (1) and Amemiya's (1974) Tobit system for Equation (2). Further, when all error correlations are equal to zeros, that is,

$$\rho_{ij} = 0 \quad \forall \quad i > j \tag{7}$$

the model reduces to one with separate probit and Tobit models from Equations (1) and (2). Tests of these restricted models against the unrestricted model can be carried out by likelihoodratio tests.

Unlike instrumental variable estimation, for which exclusion restrictions are often needed, in ML estimation the nonlinear identification criteria (i.e., linear independence of the first derivatives of the likelihood function with respect to parameters) are met, due to the functional form and distributional assumptions for the current system.

However, to avoid overburdening the nonlinear functional forms for parameter identification, we make a priori assumptions in that regard. First, while it is hard to contemplate how education per se might affect grain consumption, the educational variables are more likely to affect label use and perceived importance of grains, so they are only included in the probit equations, through which they affect consumption indirectly. Other variables used solely in the probit equations include being a meal planner and behavioral variables such as amount of exercise and being a smoker, and the attitudinal variable 'pessimistic.' On the other hand, because grain consumption is more likely to vary across regions, household types, seasons, household size, and urbanization categories than label use and perceived importance of grains, these variables are used only in the consumption equations.

To examine the marginal effects of explanatory variables, we derive relevant probabilities, conditional means, and unconditional means of the dependent variables. We partition the error vector e at the second row so that $e = [e_1', e_2']' = [u_1, u_2 \mid u_3, u_4]'$, with corresponding partitioning in the covariance matrix

$$\Sigma = \begin{bmatrix} \Sigma_{11} & \Sigma_{12} \\ \Sigma_{21} & \Sigma_{22} \end{bmatrix} \tag{8}$$

Then, by using properties of the multivariate normal distribution (Kotz et al., 2000), e_2 can be expressed, conditional on e_1, as

$$e_2 = \Sigma_{21} \Sigma_{11}^{-1} e_1 + \varepsilon \tag{9}$$

so that ε is independent of e_1 and x, $e_1 = [y_1 - z'\alpha_1, y_2 - z'\alpha_2]'$, and $\varepsilon \sim N(0, \Sigma_{22} - \Sigma_{21}\Sigma_{11}^{-1}\Sigma_{12})$.

Substituting Equation (9) into Equation (2) gives the conditional Tobit system

$$y_i = \max\{0, x'\beta_i + \gamma_{i1} y_2 + \gamma_{i2} y_2 + \varepsilon_i^*\} \tag{10}$$

Where:

$$\varepsilon^* = [\varepsilon_3, ..., \varepsilon_m]' = \Sigma_{21}\Sigma_{11}^{-1} e_1 + \varepsilon$$

$$\sim N(0, \Sigma_{21}\Sigma_{11}^{-1}\Sigma_{11}^{-1}\Sigma_{12} + \Sigma_{22} - \Sigma_{21}\Sigma_{11}^{-1}\Sigma_{12})$$

$$\equiv N(0, \Omega) \tag{11}$$

The covariance of ε^* in the second line of Equation (11) follows from the independence of ε and e_1. The motivation for writing the conditional Tobit system as Equation (10) is that parameters estimated consistently with 'unobserved heterogeneity' due to omission of $\Sigma_{21}\Sigma_{11}^{-1}e_1$. By denoting the univariate standard normal cdf as $\Phi(.)$ and the standard deviation of ε_i^* as ω_i, which is the squared root of the ith diagonal element of Ω in Equation (11), the probabilities, conditional means, and unconditional means of y_i (for $i = 1, 2$) can be expressed as

$$\Pr(y_i > 0) = \Phi[(x'\beta_i + \gamma_{i1}y_1 + \gamma_{i2}y_2)/\omega_i] \tag{12}$$

$$E(y_i | y_i > 0) = x'\beta_i + \gamma_{i1}y_1 + \gamma_{i2}y_2 + \omega_i \frac{\phi[(x'\beta_i + \gamma_{i1}y_1 + \gamma_{i2}y_2)/\omega_i]}{\Phi[(x'\beta_i + \gamma_{i1}y_1 + \gamma_{i2}y_2)/\omega_i]} \tag{13}$$

$$E(y_i) = \Phi[(x'\beta_i + \gamma_{i1}y_1 + \gamma_{i2}y_2)/\omega_i] \times (x'\beta_i + \gamma_{i1}y_1 + \gamma_{i2}y_2) + \omega_i \phi[(x'\beta_i + \gamma_{i1}y_1 + \gamma_{i2}y_2)/\omega_i] \tag{14}$$

Marginal effects of explanatory variables x, y_1, and y_2 can be derived by differentiating Equations (12), (13), and (14). The effects of each discrete variable can be calculated as the changes in these probabilities and means resulting in a finite change (e.g., from 0 to 1) in the variable, while holding all other variables constant.

In: U.S. Grain Consumption
Editors: Sara D. Torres et al.

ISBN: 978-1-61122-953-0
© 2011 Nova Science Publishers, Inc.

Chapter 2

THE CHANGING FACE OF THE U.S. GRAIN SYSTEM: DIFFERENTIATION AND IDENTITY PRESERVATION TRENDS[*]

Aziz Elbehri

ABSTRACT

This chapter examines current trends in the U.S. grain industry. Many identity preservation (IP) grain systems have emerged recently, driven by a confluence of supply and demand factors. IP grain requirements for specific production protocols, marketing channels, and quality assurance depend on whether the crops are trait-specific, non-GM (genetically modified), organic, or pharmaceutical. Cost structures vary according to the relative importance of segregation and risk management. High information management, greater market coordination, and frequent reliance on contracts characterize IP grains. IP grain markets are also inherently riskier, with volatile supply, inelastic demand, and fluctuating price premiums. Increasing grain differentiation is altering the marketing structure of the U.S. grain industry and creating possible roles for government policy, particularly in market facilitation, standard setting, and regulations affecting food safety and biosecurity.

---

[*] This is an edited, reformatted and augmented edition of a United States Department of Agriculture Economic Research Report Number 35, dated Feburary 2007.

Keywords: Identity preservation, production differentiation, specialty grain, segregation cost, traceability, quality assurance, grain attributes, risk management, information management

ACKNOWLEDGMENTS

The author gratefully acknowledges the contributions of Cullen Hawes, student intern at ERS in summer 2003, and Linwood Hoffman, William Chambers, Cathy Greene, Gerald Plato, and William Lin, ERS. Special thanks to Demcey Johnson, ERS, for managerial leadership and guidance on the project. Thanks also for very helpful reviews to Michael Boland, Kansas State University; Corinne Alexander, Purdue University; James Pritchett, Colorado State University; Warren Preston, USDA-AMS; Peter Goldsmith, University of Illinois Urbana-Champaign; David Skully, John Dunmore, and Barry Krissoff, ERS. And finally, thanks to Dale Simms and Wynnice Pointer-Napper (USDA, ERS) for editorial and design assistance.

Note: The use of brand or firm names in this publication does not imply endorsement by the U.S. Department of Agriculture.

SUMMARY

The U.S. grain system is increasingly marked by product differentiation and market segmentation. More specialty crops now require either some form of segregation or full-scale identity preservation to keep them separate from conventional commodities. Market segmentation within the grain system is driven by the need to preserve its market value, or ensure purity of the product. Internationally, U.S. grain markets must increasingly conform to a new regulatory environment reliant on traceability and identity preservation.

What Is the Issue?

Differentiated grain markets differ markedly from those for commodity grains. The commodity market is characterized by minimum common standards, a large number of buyers and sellers, and high flexibility. Price is the primary coordination mechanism, with commodity exchanges often the locus

The Changing Face of the U.S. Grain System: Differentiation... 39

of price discovery. Pricing is with reference to standard grades (e.g., number 2 yellow corn) that are broadly accepted, enhancing market fluidity. By contrast, differentiated grain markets have fewer buyers and sellers, higher costs for segregation or full-scale identity preservation, and specific quality standards, compounded by higher risks in production and marketing. Differentiated grains usually command price premiums, based on the extra costs incurred by producers and shippers and willingness to pay at the processing or retail level. This chapter examines the economics of grain differentiation, including the cost implications of different protocols, the unique risk factors of adopting IP (identity preservation) grains, the use of contracts, and the role of government as a provider of market information and facilitator of product-differentiated markets.

What Did We Find?

To preserve the identity of a specialty crop, segregation from commodity grains or oilseeds is required. In some cases, this is necessary to protect purity and to preserve the value of the specialty crop. In other cases, the goal is to prevent contamination through accidental commingling (for example, biotech or not), or to protect products that are approved only for certain uses (for example, industrial use only).

The cost structure for IP grains differs with the degree of segregation and/or IP required. For high-value grains, costs encompass both segregation and identity preservation in the supply chain, and the costs to mitigate risks specific to IP grain markets. Volume shipped, shipping method, tolerance levels, testing, and documentation requirements can influence segregation costs. Costs associated with risk mitigation depend on the type of specialty crop as well as the purity level. Lack of compliance with a product specification can lead either to a price discount or rejection of a shipment by buyers.

Price setting under an IP grain system is characterized by premiums or discounts relative to standard commodities, whether or not production and marketing is under contract. Premiums are affected by various factors, including the proximity of suppliers to buyers and the cost and availability of substitutes. For many trait-specific crops, price premiums rise or fall depending on supply conditions for the generic commodity.

Differentiated grains require more coordination between growers and handlers or processors, and more sharing of information. This arises from the trait- specific quality attributes of IP grains: within the supply chain, information must be conveyed about raw materials, key ingredients, and production/manufacturing processes. Assurance of product quality and authentication of process/product claims is often required. Farm product suppliers (for example, seed producers) must demonstrate that product attributes are verifiable and show supporting documentation.

Production contracts are important for trait-specific grain to ensure that the attribute-specific commodity is delivered and that predetermined management practices are used. For the producer, contracts can ensure a return adequate to cover costs of identity preservation and any yield drag associated with trait-specific varieties. Contracts can also ensure that there is a market for a niche product.

Production and marketing of trait-specific grains involves risks associated with price, quality, and information. Testing and documentation bring greater transparency to the transactions (in terms of quality and production processes), but the loss of anonymity also exposes producers and handlers to new risks. Farmers' management ability can affect both yield performance and proficiency with contracts and relationships. On the buyer side, contracts help meet the demand for specific product qualities, improve cost efficiencies of product processing, and reduce transaction costs.

Risks are typically higher for specialty crops than for generic crops. Non-GM crops subjected to testing run the risk of rejection. Organic grain can be accidentally contaminated. Pharmaceutical crops are not licensed for food/feed use, so risk of contact with the food supply can make their handling far more costly. Sophisticated risk management practices are required to minimize the chance of potential gene outflow. This entails a closed-loop system with rigorous quality control and a tight chain of custody.

Increasing grain differentiation in the U.S. food and feed industry may put new demands on government, but it is not clear whether USDA's traditional roles in commodity markets should be extended to specialty grains. The collecting of price information for commodities is not easily extended to specialty grains, which are heterogeneous, small in scale, and locally concentrated. Moreover, price information can be proprietary, established through private supplier-buyer contracts. Likewise, USDA-approved grades for specialty grains may not be justified since desired traits are idiosyncratic.

As differentiated grain markets expand, the U.S. grain industry faces new demands for identity preservation, segregation, and product tracing. This will require adaptations in grain production and handling, closer market coordination, more extensive information systems, new risk management tools, a better understanding of costs, and more third-party services for auditing, verification, and quality assurance.

How Was the Project Conducted?

The project was based on an extensive analysis of the literature covering industry case studies, academic research, and government documents. Key findings—especially those relating to farm risk management, cost of segregation, and IP market dynamics—are drawn directly from analyses conducted at the Economic Research Service with outside collaborators.

INTRODUCTION AND OVERVIEW

The U.S. grain system is undergoing increased product differentiation and market segmentation (figure 1). Over many decades, the production and marketing of corn and soybeans have reflected their relative homogeneity as bulk commodities used mostly for feed. While specialty crops (high-oil corn, food-grade soybeans) have long coexisted with commodities, industrial processing for the most part did not require segregation of varieties, while breeding favored input features or yield over output traits. Infrastructure and marketing were mostly in keeping with bulk-type products.

However, a number of forces—including biotechnology, industrial processing innovations, logistical advances, information and measurement technologies, and consumer preferences—have induced rapid market adjustments, creating more opportunities for differentiation and for the development of products with specific traits as farmers sought to diversify outside the commodity system. As markets responded to these economic incentives, cost-effective approaches to market segregation and identity preservation(IP) were implemented to meet the demand for trait-specific specialty grains and to capture price premiums.

Acreage of specialty crops has recently increased in the United States; the share of IP corn rose from less than 3 percent of U.S. corn in 1995 to 7.2 percent in 2003.

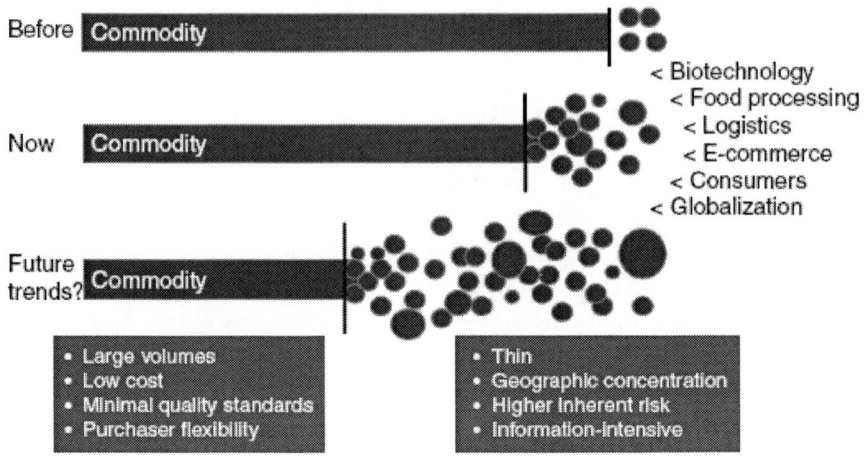

Figure 1. Grain differentation trends and grain market characteristics

However, the extent of acreage devoted to differentiated crops varies widely, from over a million acres (food-grade corn, lowtemperature-dried (LTD) corn) to fewer than 100,000 acres (high-amylose corn, short-grain rice) (table 1). The number of specialty corn types is also expanding beyond the traditional niche markets for specialized animal feed.

IP soybean acreage is hard to determine, but may range from 5 to 20 percent of total soybean acreage, depending on the extent to which sulfonylureatolerant soybeans (STS) are marketed as a specialty crop. Like corn, the number of specialty soybeans is growing, including food varieties (tofu- clear hilum) for Asian markets, soyfood products, and non-GM soybeans. Wheat and rice produced and marketed as IP have also been developed.

A number of forces affect the accelerated differentiation within grain markets. While these forces act principally at the intermediate levels of the supply chain (handling, processing and food manufacturing), consumer preferences are also at play. Consumers in industrialized countries, in response to dietary and health concerns, are more likely to demand variety- or attribute-specific grains. For example, the rising demand for low-carbohydrate food prompted greater interest in resistant starch (not easily digestible), hence offering much higher dietary fiber relative to carbohydrates. This starch, in turn, requires specialty grains that best meet specific starch requirements.

Table 1. Major specialty grains and oilseeds grown in the United States

Product type		Product name	Acreage (x 1,000 acres)	Marketing channels
Output-specific traits	Standard, with special handling	HES corn Food-grade corn Low-temperature dried corn	150 (2003) 1,200-1,500 (2002) 1,200 (2001)	Segregation within bulk
	Specialty type (nontransgenic)	High-lysine corn High-oil corn	100 (2000) 500 (2002)	Segregation within bulk
		Nutritionally dense corn	85 (2003)	
		Waxy corn	700 (2003)	
		White corn	900 (2002)	
		Waxy wheat	NA	
		"Desert" durum wheat	264 (2003)	
		Short-grain specialty rice	34.7 (2002)	
		Waxy rice	NA	
		High-oleic high-oil corn	5 (2001)	Closed loop
		Low-phytate corn	NA	
		High-amylose corn	60 (2003)	
		High-sucrose soybeans	4.5 (2003)	
		High-isoflavone soybeans	12 (2003)	
		Low-saturate soybeans	11 (2003)	
		Low-linolenic soybeans	2 (2003)	
		Blue corn	NA	Contained/ bag shipping
		Clear hilum tofu soybean	970 (1997)	
	Transgenic (GM)	High-oleic soybean	10 (2002)	Closed loop
Absence of attribute (non-GM)		Non-GM corn	300-600 (2002)	Segregation within bulk

Table 1. (Continued)

Product type		Product name	Acreage (x 1,000 acres)	Marketing channels
		Non-GM/STS soybeans	14,000 (2003)	
		Non-GM wheat	NA	
Organic	Nonconventional production	Organic corn	135 (2003)	Certification (ISO 65)[1]
		Organic soybeans	175 (2003)	
Pharmaceutical and industrial	Not approved for food/feed	Pharma crops	NA	Closed loop/ Containment
		High erudic acid ("Crambe")	8.5 (2001)	

NA = Not available.
[1] ISO 65 = ISO Guide 65 points general requirements for organic certification bodies.
Source: Various university and industry reports.

Not all product attributes are readily discernible by final consumers. "Credence" goods, like foods derived from organic grains, require process verification. By contrast, products with "search" or "experience" attributes can be distinguished either visually (e.g., clear-hilum soybeans) or through testing (waxy corn, high-extractable-starch corn). Better testing and measurement technologies have enabled food manufacturers and retailers to discern and demand attribute-specific food products and ingredients. Global supply chains are adopting new standards of safety and conformity, partly to meet trading requirements (such as sanitary and phytosanitary standards), but more often to place their proprietary consumer goods into differentiated-product markets.

Food processors' demand for trait-specific crops derives from the need to improve production and processing efficiency, reduce costs, or enhance product value. General Mills, for example, now procures variety-specific wheat and oats, relying on contracts with a network of producers and cooperatives in several States. Warburtons, in the UK, has established contracts with Canadian wheat producers to deliver variety-specific wheat.

Demand for specialty grains is reflected in buyers' willingness to pay a premium over conventional grains (see box on next page). Processors pay a premium for specialty/differentiated grains due to their production efficiencies. For farmers, such premiums must be sufficient to offset any yield drag associated with specialty grain varieties, as well as any additional costs of production or segregation.

Several supply-side forces also contribute to greater grain differentiation, including improved and novel varieties derived from biotechnology. To date, most biotech crops have involved agronomic traits (e.g., herbicide tolerance or

The Changing Face of the U.S. Grain System: Differentiation...

resistance to pests). These crops do not require segregation in the U.S. marketing system, and no market premiums apply. However, consumer aversion to biotech in some markets can lead to premiums for nonbiotech grain, pushing retail chains and agricultural suppliers to separate biotech from nonbiotech products.

The number of output-trait grains using biotechnology that have reached the market remains limited. One example is low-phytate corn, a genetically modified corn naturally high in digestible phosphorus. Hogs fed this corn excrete less phosphorus in manure, so low-phytate corn can reduce pollution from hog farms. There are several reasons for the limited number of output-trait grains from biotechnology. Much of the early effort focused on input traits, given the huge market for feed crops and expected return on investments. Concern about consumer acceptance and limited food use for some crops may have slowed the development of output-trait varieties. Nevertheless, many output-trait products are in the pipeline, with entirely new crop uses, both in food and feed (e.g., high-oleic-acid soybeans, modified-starch corn, low-phytate corn).

PRICE DETERMINATION FOR TRAIT-SPECIFIC CROPS

Basic demand and supply for commodity and trait-specific crops is represented in the figure below. Supply and demand schedules for the generic commodity are denoted by S_c and D_c. The market clears at P_c. Farm-level supply of the trait-specific crop is S_v. Demand by processors (or other end-users) at the end of the marketing chain is D_1. Farm-level (derived) demand is D_2. The vertical distance between D_1 and D_2 (price P_a minus price P_b) represents the extra cost of segregation and IP in the supply chain. Also, producers earn a premium given by P_b minus P_c.

This model identifies the additional costs associated with trait-specific crops. The supply curve for the trait-specific crop lies above that for the generic commodity, owing to higher per-unit production costs. The vertical distance between supply curves S_c and S_v represents the extra farm-level costs for supplying a trait-specific crop, while the vertical distance between D_1 and D_2 represents identity preservation costs. Given inelastic demand for the trait-specific crop, small shifts in supply (S_v) can produce large changes in producer premiums. For a sufficiently large increase in supply, the producer premium can evaporate. Moreover, shifts in supply or demand in the commodity market also affect price premiums indirectly.

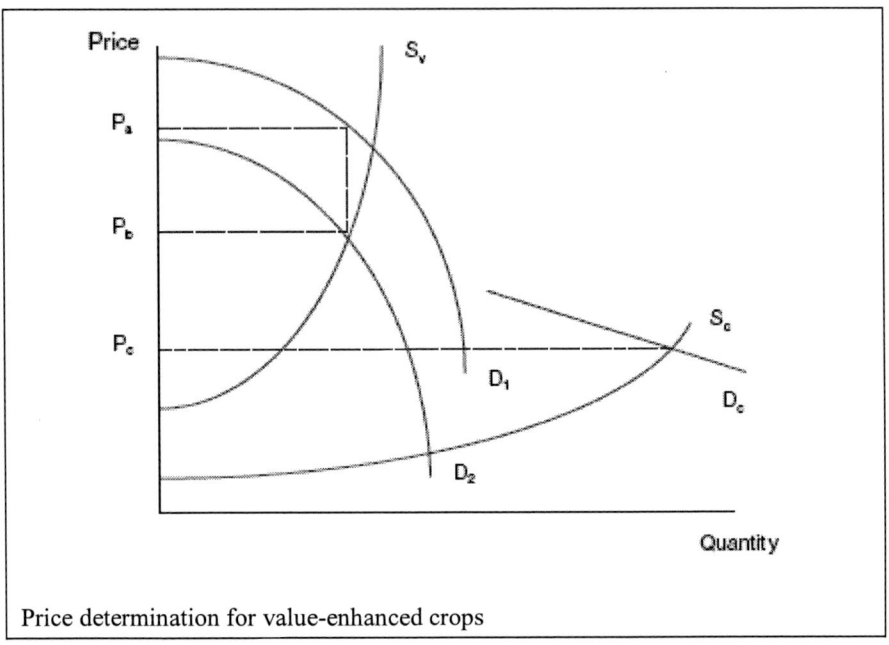

Price determination for value-enhanced crops

Biotechnology has also influenced grain markets indirectly—by yielding products (e.g., enzymes) that increase demand for specialty grains. For example, development of new genetically engineered enzymes has expanded corn processing, enabling the corn wet-milling industry to produce starches tailored to specific industrial and food uses (Hicks et al., 2002). This, in turn, has increased demand for trait-specific corn types such as waxy and high-amylase corn. Also, corn fiber is being investigated as a source of biobased industrial products, nutraceuticals, functional food ingredients, and biofuels. The current push to use crops for biofuels production, particularly from cellulose, is likely to spur the development of genetically modified crops with desirable cellulose characteristics to facilitate more efficient cellulose-to-ethanol conversion.

Innovations in transportation, logistics, and information technologies have also facilitated the marketing of differentiated grains. For example, Web-based bin-monitoring software can remotely assess the quantity and quality of grain inventories in a supplier's storage facilities. Communication networks and increased reliance on the Internet are also cutting transaction costs, especially for third-party authentication.

Differentiation is expected to grow as commodity markets respond to new economic opportunities from high-value production. Nearly 70 percent of U.S.

grain industry leaders forecast in 2002 that identity preservation would be important in 5 years (compared with 20 percent who thought it was important at the time of the survey). Likewise, 76 percent thought that diagnostic systems for quality attributes would be important in 5 years, compared with 31 percent at the time of the survey (Shipman, 2003).

Overall, differentiated grain markets differ markedly from those for commodity grains, notably in market liquidity, price premiums, cost structure, the need for segregation, diversity of standards, and the nature of and approaches to risk management. These differentiated markets are characterized by identity preservation systems, with different implications for competition and market structure than commodity markets. Consequently, the policy implications and the public role may also be different from the government's traditional role in commodity markets. In this chapter, we examine the economics of grain differentiation —the cost implications of different protocols, the unique risk factors of adopting IP grains, the use of contracts, and the role of government as a provider of market information and facilitator of product-differentiated markets.

Table 2. Grain identity preservation system: Main features and product types

	Output-specific traits	Absence of attribute (GM)	Certified organic	Pharmaceutical/ industrial (federally regulated)
Example	High-oil corn	Non-GM corn	Organic soybean	"Cystic fibrosis" corn
Production protocol	Specific variety	IP production controls Inspections	Certified organic seed Multistep inspections	IP protocols/ Isolation Dedicated equipment Field inspections
Quality assurance	Testing (Near reflectance)	Non-GM testing Auditing Certification	Non-GM testing Auditing Certification (National Organic Progam)	3rd-party audits Certification Recordkeeping Source verification
Marketing channels	Segregation within bulk	Segregation within bulk	Certification	Closed loop/ containment

MARKETING CHANNELS AND QUALITY ASSURANCE

Grain product differentiation occurs both to preserve desirable traits in a crop variety (e.g., milling yield of wheat into flour) and to exclude traits perceived as undesirable by some users, such as GM content. Differentiated grains are further distinguished from commodity grains by market prices and underlying cost structures. An identity preservation (IP) system ensures that producers, handlers, and processors keep trait-specific grain separate from other grain types throughout the supply chain. IP crops fall into four broad categories based on production requirements, specialized marketing, and quality assurance schemes (table 2).

Production Requirements and Marketing Channels

Production requirements for IP crops vary from simply using a specific variety (trait-specific crops) to specialized production methods (organic crops), from protection against GM contamination (non-GM crop) to crops that require an elaborate set of safeguards and confinement practices (pharmaceutical and industrial crops). The production requirements for IP grains are influenced by such factors as the degree of purity required, the volume shipped, and market acceptability. Marketing channels for IP crops can be separated into several types (table 3).

Segregation by Channeling

Specialty corn and soybeans produced in large volumes are segregated by channeling, which uses the existing commodity marketing system with slight modifications. Segregation by channeling flows from farm to truck/rail to elevator. The effort to minimize commingling varies from simply running equipment empty before switching varieties to designating certain days of the week or alternate sites for receiving and shipping these specialty crops. Crops handled by this method include white, waxy, yellow food-grade, high-oil, and non-GM corn; non-GM soybeans; and "desert" durum wheat (grown in the Southwest and valued for its quality attributes).

Table 3. Major marketing channels for indentity-preservation grains

Marketing channel	Channel features	Specialty crop type	Examples of IP crops
Bulk	Large volumes Blended for minimal quality Low cost/flexibility	Standard corn/soybeans	Yellow # 2 corn Traditional soybeans
Channeling	Separate sites/days/cleaning Impurity threshold: 1-5%	Enhanced value traits GM-free	White corn, high-oil corn, yellow food grade corn, waxy corn, non-GM corn
Closed loop	IP controls extend to farm Direct transfer to end-user Lower threshold: 1%	Trait-specific Limited-approval GM	High-oleic soybeans High-amylose corn Stacked GM corn
Container	Closed IP/direct end-user delivery 20-foot container/bags Lower threshold < 0.5%	Food grade Non-GM/low threshhold	Blue corn Tofu/clear hilum soybeans
Certified (ISO standards)	Certification regulations IP practices in production, handling Inspections along supply chain	Organic crops Seed production	Organic corn & soybeans Organic tofu/clear hilum Certified seed
Segregation/ Isolation	Confined production Spatial/temporal separation Zero tolerance for commingling	Industrial crops Pharmaceutical crops	High-erudic acid rapeseed "Anti-hepatitis" corn

Closed-Loop System

A closed-loop system provides more controls than mere channeling, and better protects the value of a specialty crop such as high-sucrose soybeans, high-oleic soybeans, or high-amylose corn. Production occurs almost exclusively under a contract between the grower and end-user. Typically, these production contracts mandate delivery of all production to a specified location, and require midseason inspections and return of all unused seed to the seed company. Third-party auditors also verify that the system is in fact a closed loop and that all requirements have been adhered to throughout the system.

Container-and-Bag Systems

For very small quantities of an identity-preserved grain, the container-and-bag system is an effective means of transporting and marketing. It has been used for several decades in the seed industry and in exporting tofu/clearhilum food-grade soybeans to Asian markets. With this system, 20-foot shipping containers are loaded and sealed on or near the farm where production occurs. This guarantees more stringent purity levels (< 0.5 percent of GM or foreign content) than with the closed-loop system. Specialty crops and seed marketed under the container-and-bag system are produced under direct contracts with end-users who take full charge of handling and shipping from farm to delivery. These contracts and handling arrangements normally encompass testing and tolerance/certification requirements, carried out either by the end-user or a third-party agent.

Quality Assurance

The degree of quality assurance varies widely among IP products, ranging from a simple Near Infrared Reflectance (NIR) test at the point of entry into the supply chain to a highly regulated system of verification, certification, and assurances for products under full confinement.

Whenever feasible, seed testing for specific attributes is essential for marketability of the IP product. Testing is one of the main vehicles for ensuring identity and quality of trait-specific grains and oilseeds. Sampling and testing procedures are often specified in production contracts. Testing for waxy corn types requires a single iodine-and-water test conducted by an elevator (buyer) at harvest. In many cases, a third party may also be involved in sampling/testing at selected stages along the supply chain. NIR tests have been widely used to test for protein and oil content in grains. Moreover, advances in measurement and computing, such as digital imaging (DI) techniques, are enabling tests of very small or even single-seed samples for specific variety identification.

Developments in the IP grain market and the burgeoning quality certification systems may indirectly affect the marketing of commodities, requiring an adaptation of existing grades and standards. As IP grain markets expand and their production and infrastructure become more widespread, it is likely that the grain commodity system itself would be altered by placing greater emphasis on quality (facilitated by better measurement technologies). The current grading system, which relies primarily on tests for visual traits such as cleanliness or damage, may be expanded to recognize intrinsic quality of grains and oilseeds.

When testing is not feasible (as for credence attributes), auditing, certification, and traceability systems may be needed (Dunahay, 1999). For example, organic crops rely exclusively on certification for ensuring product integrity. Organic producers are certified for observing production protocols that cover pesticides, fertilizers, cropping histories, and biotechnology. Their farms and fields are subject to inspection by certifying agencies, which are private businesses and government agencies accredited by the USDA National Organic Program (Greene and Kremen, 2003).

An IP system may or may not include source verification or full traceability, defined here as the ability to track the product backward from the point of final sale to its point of production. Full traceability is often driven by food safety management. Traceability does not affect the quality of the product; hence, identity preservation and traceability are not synonymous (see report on traceability by Golan and colleagues, 2004). The IP system for grain does not guarantee a continuous chain of custody from the final loaf of bread on the supermarket shelf back to the farm or field where the grain was grown. Under IP, testing is common at all points of transfer to verify quality, safety, or absence of GM event (attribute) or quality trait, but the specific grain origin (individual farmer) is lost.

The current U.S. grain market does not observe full traceability of grains from field to shelf. Nevertheless, the regulatory environment is changing significantly. The U.S. Public Health Security and Bioterrorism Preparedness and Response Act of 2002, the European Union's traceability and labeling directives beginning in April 2004, and the Biosafety Protocol on Biodiversity starting in 2006 are all inducing agribusinesses, particularly those at the start of the supply chain such as grain elevators and grain warehouses, to track shipments and supplies more than before. These regulations are expected to require that food and feed manufacturers, processors, transporters, and importers keep records of the (immediately previous) source of food, feed, or ingredient being accepted; the transporter who delivered it; the next recipient of the firm's product; and the transporter delivering it to that recipient. For grain markets, this partial traceability approach can lead back to the food/feed ingredient supplier or a group of elevators/farmers in the event of a food recall (Farm Foundation, 2004). A fully traceable system, as exists in the seed industry,[1] can develop only if there are strong economic incentives, such as foreign buyers with specific food safety concerns who are willing to pay very high premiums.

E-Commerce and Third-Party Services

The use of IP protocols requires that critical steps in production, shipping, and processing be observed and documented and that transaction information be managed. As a result, databases including information-tracking software are critical. The Internet can lower transaction costs managed by third-party data service providers.[2] A number of Internet-based service organizations have emerged to help facilitate the marketing of specialty grains and oilseeds, including testing/certification, matching producers with customers, information tracking, and e-commerce services.[3]

COSTS OF SEGREGATION

Technological, processing, and logistical innovations have created opportunities for specialty products that command premium prices. But an identity preservation market is predicated on farmers willing to grow such products and handlers to market them to end-users at acceptable costs. To earn the IP premium, those involved must incur added segregation costs and bear the risks inherent in IP production and marketing.

The cost structure for identity-preserved crops differs from that of commodity crops in two important ways:

(1) the added costs of segregation and identity preservation in the grain supply chain; and
(2) the costs to mitigate risks specific to IP grain markets, risks rising from the characteristics of differentiated-grain markets, and the specific requirements for production and marketing.

Sources of Segregation Costs

IP costs derive from the volume of grain handled, levels of purity required, handling infrastructure, and the extent of risk and risk sharing. The relative impact of these factors on IP costs also differs by the type of specialty grain, and can vary at each stage of the supply chain.

Basic IP costs involve the physical separation of grain, including dedicated storage, handling, and transport of harvested products. The main sources of added costs over conventional varieties are seed (including technology fees) and special transportation, handling/drying, storage/segregation, and management. (Farm surveys show that these segregation costs differ widely among specialty grains.) For high-oil corn, the largest additional cost is seed, as proprietary seed commands a technology fee, while the ease of testing for oil level makes physical segregation unnecessary (Fulton et al., 2003). For waxy, white, or food-grade corn, IP costs are more substantial for physical segregation (transportation, handling and drying, and storage). For specialty soybeans like tofu or seed soybeans, the main additional costs are seed technology fees and transportation.[4]

IP costs are also affected by whether verification claims are required. Testing or documentation requirements are particularly important for crops marketed as non-GM, which require additional steps to avoid accidental commingling on the farm. In practice, this means growing border fields and staggering harvests to avoid pollen drift and contamination from non-GM fields. In the long run, farmers who choose to produce differentiated grain must also incur fixed costs in buying storage bins and drying/harvesting equipment.

Grain elevators typically store and mix grain of varied quality and grades, in response to market signals. Identity preservation and segregation preclude these practices. Additional costs derive from physical separation, including separate storage and identity verification.[5] Marketing costs of IP products include expenses associated with contract negotiation and the procurement of suitable grain or oilseeds from farmers.

Hidden or opportunity costs for IP elevators include loss of margins from forgone opportunities and from underutilized storage capacity, which are significant for some elevators (Bullock et al., 2000; Qasmi et al., 2004). Maltsbarger and Kalaitzandonakes analyzed segregation options for high-oil corn at the elevator stage and broke out separation costs into coordination, logistical, and indirect categories. They found that the average IP cost varies widely with volume handled and that the relationship between IP cost and IP volume is determined mostly by the physical configuration of the storage facilities. Indirect costs account for over half the total segregation cost and are higher with larger volumes, highlighting the loss of efficiency and revenue to accommodate IP inflows. Other studies have examined the indirect costs of delayed deliveries and loss of flexibility in moving large volumes of grains during the peak harvest period.[6]

Variation in Cost Factors

The impact of **volume marketed** on IP costs depends on the grain-handling infrastructure. Elevators that are able to segregate most effectively have many bins of varying capacity, as well as multiple pits and elevator bucket legs; these features enable the elevator to dedicate units to specialty grain, reducing the likelihood of commingling (Herrman et al., 2002). The volume of IP grain within the infrastructure is significant in selecting the optimal segregation strategy, including whether to designate elevators as IP-only facilities. For grains and oilseeds in Indiana, low-volume flows make it cost effective to designate IP facilities only when the processor of the product is local (Vanderburg et al., 2003). In IP-dedicated plants, the increased transportation and handling costs (of getting more specialty grain from farther away) are more than offset by the elimination of segregation costs.

The method of grain shipment also influences IP costs. Containerized IP shipments are becoming the preferred response to the growing demand for segregated specialty grain (Reichert and Vachal, 2000). While bulk systems (train/barge) have historically been a cheaper way to move grain because of economies of scale, containerized shipping reduces time in transit. This is particularly important to customers requesting just-in-time service. Other advantages include better inventory management along the supply chain and better matching of supply with demand (and hence lower seasonal price fluctuations).

The **tolerance level for impurities** allowed in grain also affects IP costs. The higher the degree of purity required, the higher the cost to validate compliance (table 4), especially for non-GM grain (Giannakas and Kalaitzandonakes, 2005). Segregating grain into biotech and nonbiotech entails greater marketing risks than trait-specific grain such as high-oil corn.[7] Under generous threshold requirements (95 percent or higher product purity), segregation costs are manageable within the current handling infrastructure (Lin et al., 2000). At higher purity thresholds (99 percent or higher), production and marketing of specialty crops can add significantly to IP costs.

Table 4. Estimated costs of segregation at a grain-handling facility from previous studies

Author(s)	Commodity	Year	Volume handled (bu.)	IP system	Costs ($/bu.)
Herrman,	Wheat	1999	6,500/hr. (model A)[1]	High-quality HRW	0.053-0.056
Boland, and Heishmann			7,500/hr-10,000/hr (model B)[2]	High-quality HRW	0.023-0.032
Bender, Hill, Wenzel, and Hornbaker	Corn and soybeans	1998	n.a.	High-oil corn STS soybeans	0.06 0.18
Good, Bender, and Hill	Corn and soybeans	1999	n.a.	Non-GM corn Non-GM soybeans STS soybeans	0.01 0.078 0.117
Smyth and Phillips[3]	Canola	1999	n.a.	Non-GM canola	0.6(1% tolerance level)
Maltsbarger and Kalaitzandonakes	Corn	2000	200,000 to 500,000 during peak harvest	High-oil corn	0.164-0.274
Wilson and Dahl[3]	Wheat	2002	n.a.	Non-GM wheat	0.01 45 (5% tolerance) 0.0336 (1% tolerance)

[1] Model A: the elevator is characterized by 1 drive, 1 elevator bucket leg, and 2 dump pits.
[2] Model B: the elevator is characterized by 1 drive, 2 elevator bucket legs, and 2 dump pits.
[3] Costs of segregation entail those from growers to those from either importers or domestic end-users.
n.a. = Not available,

Table 5. Producer risk in specialty grain: A typology

Risk type	Risk category	Risk common with commodity	Definition	Sources of risk
Base price	Market	Yes	Risk of lower-than-expected grain prices other than changes in expected market price premiums	
Price premium	Market	No	Risk of a change in premium without a change in quality within the crop year	Risk in producing value-enhanced grain (VEG) in open market
Market access	Market	No	Risk of not having a viable market for the crop • Short run: for either a VEG crop not grown under contract or the overproduction of a VEG crop • Long run: risk of VEG disappearing from market following specific investments	
Quality	Production	No	Risk of an unexpected quality level in the grain that affects the grain's value through discounts or reduced premiums Risk of rejection from low quality	Chemical composition and test weights; contamination risk (GMO) Storage quality risk post-harvest; measurement risk (testing)
Yield	Production	Yes	Risk of lower-than-expected production (different from yield drag)	Weather conditions, variety, unknown yield drag, soil fertility, pest pressure, and timing of field operations

Table 5. (Continued)

Risk type	Risk category	Risk common with commodity	Definition	Sources of risk
Contract	Business	No	Risk of contract default by producer/contractor	Default during crop year from lack of performance; termination of multiyear contract after 1 year; risk of nonpayment upon delivery
Relationship	Business	No	Risk of adversely affecting critical relationships with buyers, suppliers, or other resource providers (unique to VEG; determined by contracts or outside contracts)	Losing access to landlord/lender, supplies, technology, knowledge, and markets
Product liability	Business	No	Risk that a producer will be liable for problems	Contamination with GMO or food safety; with grain liability specified in contracts; GMO contamination in organic will prevent organic labeling
Investment	Financial	No	Risk associated with returns on a long-term asset	Variability in returns (annual changes in costs and revenues); loss of the asset (fire, theft, natural disaster); uncertain long-term returns on investments

Source: Bard et al. (2003).

Under a stricter tolerance level (99 percent or higher purity), the high risk of rejection induces greater testing costs. IP costs are incurred in both the commodity and IP systems when the IP product is either unapproved or unacceptable in some markets (i.e., Starlink corn variety approved for feed but not food in the United States). By contrast, fully acceptable IP products (i.e., GM crops approved in all types of uses and all markets) incur costs only in the IP system. The smaller the tolerance or acceptance of a product in a market, the greater will be the effort—and cost—to maintain perfect isolation.

RISK MANAGEMENT

The IP cost structure is also contingent on managing risks inherent in IP markets. These risks can arise from pricing factors (price premiums, quality, and information), production contracts, longrun investments, and farmers' management ability, which affects both yield performance and proficiency with contracts and relationships. The nature and scope of risk vary depending on the type of IP crop. These risks are examined for three categories of IP crops: trait-specific, non-GM, and pharmaceutical/industrial.

Risks Associated with Trait-Specific Crops

Farmers face several risks when they grow trait-specific grains. These fall under four broad categories: market, production, business, and financial risks (table 5).

- **Market risk** arises either from uncertainty in finding buyers or rejection of shipments if specific grain standards and characteristics are not met. Market-type risks include base price risk common with commodity crops and price premium risk specific to IP crops.
- **Production risk** includes both yield and quality risk. The quality risk may arise from inadvertent commingling of grains with different characteristics or from unfavorable weather. Deviation from pre-specified quality may result in lower premiums, discounts, or outright rejection by the buyer.
- **Business risk** includes possible contract default by producers or contractors, as well as potential liability for any problems that arise

with the grain. Critical relationships may also be strained or broken under specialty grain production.
- **Financial risk** is associated with investment risks due to variability in returns and loss of the asset. Trait-specific grain production may include investment risk above that expected for traditional commodities due to specialized equipment or facilities. Long-term returns on these investments may be uncertain since production contracts are typically for a single year. If the producer loses the contract or if the economics of the product become less favorable, the returns on the investment may be reduced.

Bard and colleagues (2003) ranked these risks using an Illinois farmer survey and found that the top three risks faced by IP producers are related to price premiums (39 percent of respondents), yield (25 percent), and quality (22 percent). The key factor that draws farmers into and out of specialty corn is the price premium.

Farm surveys conducted by the U.S. Grains Council (1996-2001) show a high degree of entry and exit into and out of specialty crop production each year (Stewart, 2003), as much as 30 percent in the case of corn (figure 2). The decision to enter or exit the specialty grain market may be linked to yield performance (figure 3). Exiting farms may be either poor production managers or have unsuitable land or growing conditions. The high degree of entry and exit mirrors the higher fluctuations of supply and demand in differentiated grain markets.

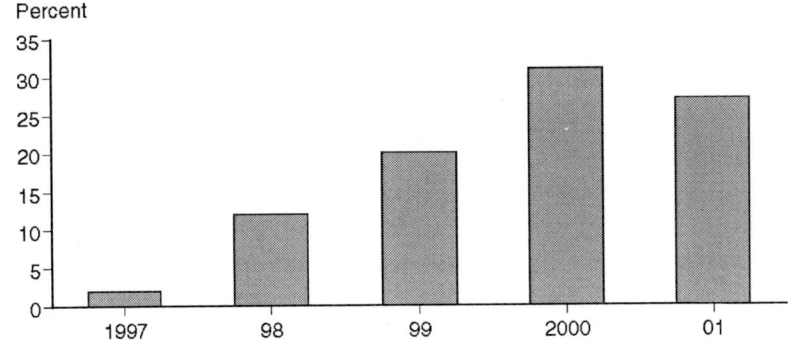

Source: U.S. Grain Council.

Figure 2. Share of corn growers exiting specialty grain markets, 1997-2001

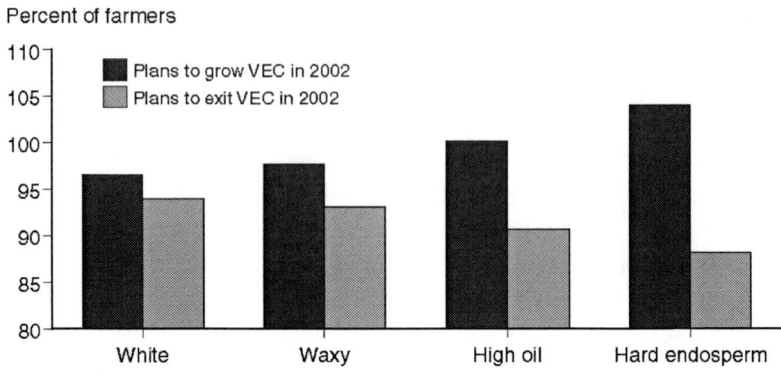

Source: U.S. Grain Council.

Figure 3. Specialty corn yields for farmers who maintain versus those who exit value-enhanced corn (VEC) production, 2002

Risks Specific to Non-GM and Organic Crops

Non-GM crops are subject to the added risk of testing for purity, which could result in rejection of shipments. The risk level depends on the testing technology and the purity threshold required. The less accurate the testing or the higher the purity threshold, the greater the risk.[8] This is a greater risk than with trait-specific grains, which incur only price discounts for inferior quality. Buyers risk receiving non-GM crops accidentally commingled with GM varieties, and so specify testing/segregation methods through contract terms, test specifications, and penalties (Wilson et al., 2005).

Organic grain producers face risk from accidental commingling with GM crops since organic regulations prohibit genetic modification. In addition, the threshold for purity in organic crops can be more stringent than for non- GM IP crops. These stringent private standards are common for exports to Europe or Japan, where some buyers may demand near-zero tolerance, while others may allow small amounts of impurities (generally above 99.5- percent purity thresholds). Organic farmers use numerous management strategies—including buffer zones, careful timing of crop planting, and crop monitoring—to minimize the possibility of accidental contamination, but the effectiveness of these management strategies varies by crop. Self- pollinated crops such as soybeans or barley pose less of a problem than open- or wind-pollinated crops like corn.

Risks Associated with Pharmaceutical and Industrial Crops

Unlike other genetically engineered crops grown for food or feed uses, pharmaceutical crops require a production license from USDA/APHIS, and must include a containment plan for pharmaceutical plants during their production, handling, and movement in and out of the field. APHIS reviews and pre-approves all plans for seed production, timing of pollination, harvest, residue destruction, shipment, confinement, and the storage and use of equipment. Field inspections can take place up to five times during the growing season, coinciding with critical periods of the growing season. Farmers contracting with biotech companies that hold an APHIS license are required to undergo training in license requirements and implementation.

Risk management drives the cost structure for pharmaceutical crops. Sophisticated production and handling aim to (1) contain the potential gene outflow and impacts on nontarget organisms, as well as workers' health; (2) create a tight closed-loop system to minimize any possibility of commingling with the food supply; and (3) create a set of quality control procedures with a tight chain of custody to satisfy the isolation and confinement requirement. Given the potential risks and liabilities associated with accidental commingling of pharmaceuticals with food grains, and with the daunting task of insuring the 100-percent containment requirement, food and biotech industries have taken a precautionary approach to pharmaceutical crops and risk assessment-based regulations (Elbehri, 2005).[9]

INFORMATION INTENSITY AND MARKET COORDINATION

Key to successful risk management for identity-preserved crops is the ability to manage the abundance of information required as the product moves along the supply chain. The need for more information arises from the high specificity of quality attributes in IP grains. Quality traits of IP crops fall into several categories:

(1) Increased levels of one or more of the desirable components in grain (protein, oil, starch, minerals);
(2) Absence or reduction of a problematic component of grain (phytate, linoleic oil);
(3) Modifications to include a desirable or specialized composition;

(4) Ingredients not previously sourced from plants (pharmaceuticals); and
(5) Modifications to generate completely novel crops (nutraceuticals).

Information Intensity

The need for more detailed information concerning grain-based food ingredients forces all agents in the supply chain to monitor quality and convey information about raw materials, key ingredients, and production/manufacturing processes, and to provide assurance of product quality and authentication of process/product claims. Farm product suppliers must demonstrate that product attributes are verifiable and show supporting documentation. For example, non-GM soybeans (other than sulfonylurea-tolerant soybean varieties) require documentation showing when the tests are required by food manufacturers and validation test results. Non-GM soybean ingredients in food products must also be accompanied by third-party certification that demonstrates their non-GM origin.

With increased differentiation, information flows become critical to mitigate risk and capture higher market value. Information management thus enters into the cost structure of IP production. Testing and third-party certification add to transaction costs. Moreover, information adds to the riskiness of the IP market. Compared to the highly efficient traditional commodity system, the differentiated information-intensive system is more transparent as more information accompanies the grain shipment. But this increased transparency brings new risks in the form of liability, intellectual property protection, performance accountability, and business relationships (Beurskens, 2003).

Market Coordination: Role of Contracting

The more information required during transactions along the supply chain, the greater the need for market coordination. Traditionally, less perishable products such as grains, oilseeds, and cotton (with the exception of niche markets) have relied mostly on open or spot markets. Storage and buffer stocks have facilitated vertical coordination and hedged against spikes in supply and demand. For grain commodities, spot markets have been sufficient for price discovery, as price and standard grades capture all the information required by buyers and sellers. Under this system, generic commodity cash markets and forward contracts (which specify volume and price only) have sufficed.

However, with the growth of information-intensive IP grain and oilseed markets, contracts have become more important (table 6). The rules of competition are being redefined between suppliers and customers, as well as between customers and among competitors. The greater the information flow along the supply chain, the greater the need for production contracts.[10]

Producers use contracts to ensure compensation for additional IP costs and any yield drag associated with trait-specific varieties—and to guarantee a market for the niche product. The producer is thus able to reduce financial and marketing risk, access new technologies and markets, and lock in price premiums.[11] Buyers use contracts to help meet the demand for specific product qualities (including food safety), improve cost efficiencies of product processing, and reduce transaction costs (Jackson and Cuppy, 2000). The buyer is thus able to maintain quality control and manage supply. In some cases, contracting also meets a need to protect intellectual property.

Table 6. Benefits and risks of contracting specialty grains

Producer benefits	Contractor benefits
Reduced financial risk Access to new technologies Access to new markets Price premium Reduced marketing risks	Quality control and supply management Reduced financial risk Control of technology and markets
Producer risks	**Contractor risks**
Long-term investments and short-term contracts Payment risk Limited returns Reduced management control Identity-preservation requirements	Finding amenable parties Litigation possibilities Control over technology Producer reliability

Source: Jackson and Cuppy (2000).

Table 7. Specialty grain contract provisions and less desirable aspects of contracts (Indiana farm survey)

Contract provisions (% respondents)	Least desirable aspects of contracts (% respondents)
Delivery to specific location (89)	Delivery date unknown (49)
Delivery on specific dates (74)	Delivery location (33)
Plant variety from designated list (71)	Additional costs (30)
Store crop on farm (71)	Yield penalty (27)
Provide samples for quality testing (42)	Quality standard (27)
Specific pricing method (e.g., forward contracts) (40)	Identity preservation (25)
Specific pricing window (e.g., Sept.-Jan. only) (37)	Loss of control (22)
More intensive production management (31)	Timing of payment (15)
Specific handling equipment and instructions (29)	Additional investment (9)
Specific harvesting equipment or technique (27)	Input requirements (9)

Source: Fulton, Pritchett, and Pederson (2003).

Production Contracts: Specifications and Types

Contracts for IP grains are typically for one season and are contracted between farmers on one hand and seed suppliers, handlers, intermediary firms, and processors on the other. Most contracts stipulate a specific variety, delivery time, delivery place, and dedicated storage of the crop on the farm (table 7). Specifications for delivery locations (in 89 percent of contracts) and dates (in 74 percent) were the most common requirements in both production and marketing contracts. In production contracts, quality control is handled through variety specification (71 percent of contracts) or through sampling and quality testing (42 percent).

Contracts may also differ in how they handle property rights. Contracts between farmers and seed suppliers preserve seed developers' intellectual property rights for new varieties. An example is DuPont Optimum seed for high-oil corn, used as an input to livestock feed. High-oil corn meeting certain standards earns a price premium in the market. Developers of seeds with higher oil content seek to maintain their property rights and generate rent in the form of a technology fee. The innovating firm (DuPont) grants a license to seed companies. The contract specifies a premium and requires growers to provide evidence of applying specific inputs.

Contracts also differ in how they are enforced in case of a breach or lawsuit (Sporleder and Schmidt, 2003). Some contracts call for penalties for noncompliance or even indemnification of the buyer. Others have process and quantity specifications, but failure to comply involves no legal liability. For example, organic contracts in Illinois are highly specific and third-party verified, but involve no legal liability for failure to deliver. In the case of poor performance, producers forgo the premium and are dropped from a list of select suppliers. The most common contracts specify minimal management processes (variety/hybrid and quantity) and do not require third-party verification.

Risk management is one of the main motivations behind contracting. However, contracts also bring some risks of their own (Bard et al., 2003). Among these, the failure to produce to contract standards will result in loss of a contract's premium rates, or nonrenewal/termination of the contract (Hayenga and Kalaitzandonakes, 1999). Under contracts, farmers might feel a loss of independence in submitting to certain terms (permitting field inspections by buyers or designated third-party certifiers, applying certain production practices, planting specified varieties, etc.).

Factors Affecting Contract Use and Frequency

Contracts are more likely to emerge for products with attributes that are difficult or expensive to measure than for products that have easily verified attributes and that can be traded more efficiently via open or spot markets (Hayenga and Kalaitzandonakes, 1999; Chambers and King, 2002). STS soybeans, for example, had been grown under contract up to 2001 as a distinctly non-GM variety. However, development of easier and cheaper testing for GM content in soybeans has lessened the allure of STS soybeans, which have become largely "commoditized" since 2001.

A second factor affecting frequency of contracts is the specificity of product use and the willingness of contractors (buyers) to pay a premium for the specialty grain or soybeans. Processors desirous of high quality and purity levels will pursue very structured contracts. Maintaining ownership of the seed and crop is a strategy that many companies use to protect proprietary intellectual rights. Specialty crops that command high premiums, such as high-amylose corn, have 100 percent of their acreage under contracts.

Also affecting the frequency of contracts is ease of entry. In the case of organic grains, most are not grown under contracts because the organic grain market is contestable, with numerous global suppliers willing to produce and sell these grains (Ginder et al., 2000a). This may also be facilitated by the existence of a standard USDA label verifying that products are organic.

Given the diversity of factors affecting recourse to contracts, it is not surprising that the frequency of contracts will vary greatly between IP crops. A farm survey by Bender et al. (2000) showed that specialty grains purchased via farm contracts ranged from 71 percent for food corn to 80 percent for food soybeans and 96 percent for high-oil corn. Good and Bender (2001) found in a survey of corn and soybean handlers in Illinois that the share contracted varied from 9 percent for non-GM corn up to 95 percent for high-oil corn. The low rate for non-GM corn is partly due to the ready availability of corn that can be tested cheaply for non-GM content. The protection of intellectual property rights explains the frequent contracting for high-oil corn (despite readily available technology to identify high oil content).

Effect of Contracting on Price Risk and Risk Sharing

The two most commonly cited reasons for entering into contracts are managing risk and minimizing production and/or transaction costs (Ahearn et al., 2003). If either party at the time of contracting knows the value of the product, or if there is unequal information available to contracting parties, risk aversion may lead to risk sharing. Production contracts for specialty grains determine how risk is shared between producers and buyers. Establishing a price or price formula at contract time can protect both sides from adverse price changes. However, different price formulas are used.

Ginder et al. (2000a; 2000b) identified three common types of contracts, based on the method by which the specialty grain price is determined (table 8). The most common is *market price plus a premium*. This leaves the farmer with all the yield and price risk associated with commodity production, but adds a fixed premium to cover the additional costs of specialty production. This type of contract is typically used with relatively high-volume specialty varieties that exhibit a small yield drag. Examples include high-oil corn, low-temperature dried corn, white corn, and waxy corn.

Table 8. Types of contract by specialty corn and soybean variety

Crop	Market price plus premium	Flat price per bushel	Flat payment per acre	Combination of premium methods	Other premium method
Percent of contracts					
Corn					
High oil	76	6	2	8	8
Waxy	62	14	*	*	24
White	53	35	*	6	6
Yellow food grade	33	67	*	*	*
Non-GM	33	33	*	*	33
Organic and pesticide free	*	100	*	*	*
Soybean					
Tofu or clear hilum	73	10	3	7	7
Organic and pesticide free	20	70	*	*	10
STS	61	28	6	*	*
Non-GM	36	45	*	*	19

May not sum to 100 percent because of rounding. * = less than 1 percent.
Source: Ginder et al. (2000a and b).

The second contract type stipulates a *flat price per bushel* produced. These contracts place the market price risk with the buyer while the farmer retains the yield risk. Flat-price-per-bushel contracts have been used in situations where there is a large yield drag, as with organic or certain specialized hybrids. For organic grains, yields may not always be substantially lower than for conventional grains. However, Ginder et al. (2000a; 2000b) found that organic corn (and soybean) yields were about 30 (25) percent lower than the corresponding average yield for the surveyed specialty (non- organic) crops. Flat-price-per-bushel contracts can be effective because they give farmers extra incentive to improve yields.

The third contract type stipulates a *flat price per acre*, which provides farmers a fixed payment regardless of the market price or yield situation. The yield and price risks are borne by the buyer. In this case, the buyer-contractor provides the seed and holds legal ownership of the growing crop through harvest, when the entire crop is delivered to the contractor at the agreed price.

Such contracts were a very small portion of those surveyed, possibly due to price support programs and farmer perceptions that they limit autonomy. Buyers also face high risk with these contracts and may use them only with strict management requirements for the farmers. Moreover, flat-price-per-acre contracts are offered only in geographically specific areas with relatively low yield risk. Specialty grains and oilseeds grown under these contracts (high-amylose corn, low-saturate soybeans, and high-sucrose soybeans) tend to have lower volume and expected yields. Hence, paying producers per acre of production is a way to transfer risk from the more risk-averse producers to the less risk-averse buyers when yield drag is high enough to discourage production.

PRICE AND MARKET DYNAMICS

In IP grain and oilseed markets, premiums can vary significantly among farms for the same crop, due both to differences in quality and in farmers' negotiating abilities. Fulton et al. (2003) reported that median price premiums for specialty corn and soybeans ranged from $0.13 to $0.36 (5.37 to 14.87 percent) per bushel. For some specialty corn types (waxy and highamylose corn) that command high price premiums, direct contract negotiations between growers and corn processors reflect unequal bargaining power, affecting the price premium received by growers. Within this vertically coordinated structure, the price-affecting power (oligopsonistic behavior) of processors may dampen prices and farmer returns (Elbehri and Paarlberg, 2003).

Annual swings in price premiums are related to higher variability in supply and demand for IP than for conventional grains and oilseeds. For many specialty grains, price premiums go through cycles. At first, premiums are high as buyers entice producers to enter into the production process. Prices then decline as more producers enter the market, until price premiums stabilize. For example, high-oil corn acreage expanded in the 1990s, reaching 1 million acres in 1999, before falling to 600,000 acres in 2001 and then stabilizing. Lower premiums and reduced demand due to substitution of less costly fats and oils in livestock feed were behind the drop in acreage.

Excess supply also erodes price premiums, particularly for non-GM crops. For example, non-GM corn has seen its premiums swing substantially due to large supply and demand imbalances between years. Unlike trait-specific specialty grains, non-GM grains can experience market conditions where price premiums fail to emerge, irrespective of the added IP costs involved in segregation. Moss et al. (2002) showed that one reason price premiums for non-GM corn are small—relative to IP costs—is an excess of non-GM corn after food-corn demand is met. Non-GM corn experienced a sharp decline in its price premium from 2000 to 2001, partly due to an oversupply, but also to the ease of testing for GM presence. In recent years, as corn prices have increased due to rising biofuels demand, it would take significant price premiums to lure corn producers to specialty corn.

THE ROLE OF GOVERNMENT IN DIFFERENTIATED GRAIN MARKETS

Since accelerating grain differentiation reflects the market response to changing consumer preferences, technological advances, and increased globalization, what role(s) should the government play. Is there a clear "public good" or "efficient market" argument for government intervention. Are there identifiable market externalities (either negative or positive) that would cause social costs or benefits to diverge from private ones and would, therefore, justify public intervention.

Under the "efficient market" argument, the government role would be market facilitation. How would this apply to IP grains characterized by small size, low liquidity, and undeveloped open pricing mechanisms. One government role currently applied to commodities is price collection. Prices quoted for standard grades, collected and published by USDA's Agricultural Marketing Service (AMS), are, however, of limited use to producers or traders of trait-specific crops. Price information for differentiated crops is more difficult, or expensive, to obtain. Moreover, it is not clear whether premiums or discounts should be reported on a real-time basis or based on periodic surveys. More important, there is no clear "efficient market" argument for public disclosure of private contract terms.

Similar issues arise in the context of insurance programs. Current insurance policies designed for commodity crops are based on market prices, historical yields, and public grade standards. Adapting these insurance programs for specialty grains would require additional information on expected price premiums or discounts, quality traits beyond minimum grade standards, and expected yield differentials. The lack of universal quality standards for specialty grains, and the unavailability of publicly collected prices, may render the extension of crop insurance to specialty crops very difficult.

Currently, most available information on specialty grain markets has come from university or private industry surveys. For large specialty grain markets, direct public provision of national market information, even if desirable from the "efficient market" argument, may be less effective than through a third-party provider (university or trade association). For example, USDA-AMS has contracted with Iowa State, North Carolina State, and Cornell Universities to survey organic grain producers and with (seed) dealers in Midwestern States to ascertain planting and harvesting intentions beginning in 2003.

Another government function currently in use for commodity markets that may not be easily extended to specialty grains is the establishment of standards for grain grades and quality. The grading system for grains and oilseeds has served the homogeneous market well for many decades. These standards specify USDA-approved sampling, inspection, and measurement procedures that are well accepted, quick, and relatively inexpensive. However, this approach may not be entirely suitable for differentiated grain markets with product-specific traits and attributes. Testing for value- enhanced crops requires using genetic markers to identify specific varieties and tests to verify the presence of added or altered traits or nutritional properties. Currently, private standards and contractual specifications are used in IP markets to meet desired attributes. While some harmonizing of these private standards might be warranted, standards themselves are privately decided between sellers and buyers, and direct government intervention may be neither justified nor demanded by market agents.

The "efficient market" argument for government role is more supportable in harmonizing and standardizing test methodologies, ensuring consistent and reliable measurements within the IP grain system. The current function of USDA's Grain Inspection, Packers and Stockyards Administration (GIPSA)— certification and harmonization of measurement technology— can be extended to differentiated grain markets. USDA is standardizing testing methodologies, evaluating testing and laboratory services, and developing new testing and analytical methods for end-use quality attributes. This not only facilitates domestic markets, but can help gain acceptance of U.S. products by foreign buyers. USDA's Process Verification Program is another example of market facilitation, applying internationally recognized standards to certify private firms' claims about quality control. This helps producers, marketers, suppliers, and processors to assure customers of their processes to provide consistent quality products.

An example of the "efficient market" argument justifying a public role is USDA's National Organic Certification Program begun in 2002, which provides consistent labeling of organic products. This program simplifies consumers' choice (through a clearly identifiable USDA organic label) and safeguards producers who abide by a specific production protocol in return for likely price premiums.

The "public good" argument in favor of public intervention in differentiated grain markets is evident with the need to prevent disruptions to food markets and/or to provide safety assurances to the public. This need translates into government regulations at the marketing or even production stage. As an example, the production and processing of plant-made pharmaceuticals are directly regulated via licensing by both USDA and FDA. In this case, regulation both safeguards potentially large economic and health benefits from plant-based pharmaceuticals and protects against potential liabilities and market disruptions due to inadvertent contamination. Under this tight regulatory environment, the growth of these crops will depend on strong and adaptable regulatory oversight, along with technological solutions to the containment challenge (Elbehri, 2005).

Another public imperative is national security. Here, the 2002 U.S. Bioterrorism and Biosecurity Act, intended to safeguard the U.S. food system against accidental or terrorist attacks, falls under the "public good" argument. This regulation requires "step-back/step-forward" traceability for all food and feed moving within commercial channels and encompasses both commodity and IP grain. Implementation of these traceability requirements is likely to stimulate IP grain markets by further justifying data tracking and information infrastructure along the supply chain.

The government role in harmonizing standards, labeling, and tolerances also extends into the international arena, the source of most demand for U.S. IP grains. The international regulatory environment is also changing rapidly.

Among the recent changes is the new European Union food traceability and labeling law, and the implementation of the Cartagena Protocol on Biosafety.[12] All these developments require more traceability and identity preservation, heightening the need for harmonization and recognition of mutual standards. The latter can be achieved through negotiations—whether through multilateral standards-setting organizations such as Codex, bilateral negotiations with economic partners (such as the European Union), or within the context of regional free trade agreements.

CONCLUSIONS

The trend toward more differentiated grains is a result of economic, technological, and structural forces. This trend has accelerated recently with an increasing number of specialty grain markets requiring identity preservation systems with separate marketing channels. Identity preservation is driven by the need to protect either the purity, and thus the value, of the specialty crop itself or the main (i.e., non-GM) commodity from contamination through accidental commingling. A market for identity-preserved products arises when buyers are willing to pay more for a trait-specific product and when farmers and handlers respond to market premiums.

Biotechnology is one of the major drivers of grain differentiation. Genetic engineering has enabled end-use applications for crops with specific attributes. For example, low-phytate corn, a genetically modified corn naturally high in digestible phosphorus, generates less phosphorus in manure and hence lowers pollution from hog farms. Biotechnology has also facilitated other innovations, such as industrial processing of crops and enzyme advances, which enable greater differentiation and increase the demand for specialty crops. Innovations in logistics/transportation and structural changes in the retail industry are also enabling the development and marketing of differentiated products, cutting transaction costs and making the information-intensive systems required economically feasible.

Communication networks and the Internet have introduced buyers to agricultural products with specific traits and allowed them to verify actual characteristics against product claims. Moreover, consumers in high-income countries are demanding more specific products, motivated by changing dietary and health concerns, concerns for food safety, and social or ethical considerations. The ability to meet these consumer demands is enhanced by upstream innovations in the food industry. For example, increased demand for low-carbohydrate food signals demand for fiber-rich grains, which is met by technological innovations in specialty starches and the crops that provide them (e.g., modified-starch corn).

The cost structure for identity-preserved crops differs from commodity grains and includes both the added costs of segregation and the costs to mitigate risks specific to IP grain markets. The risks derive from one or more pricing factors (price premiums, quality, and information) and production contracts, which are more prevalent in IP than conventional grains. Contracts specify production protocols to ensure IP. The nature and scope of these protocols increase in complexity depending on the type of IP crop, growing more complex from trait-enhanced specialty grains to non-GM crops, to organic grains, and finally to pharmaceutical/industrial crops that are not approved for food or feed use.

At the producer level, segregation costs may include specialized storage and transportation, or measures to prevent accidental commingling with GM crops. Growers of IP grains face much larger price swings (from the more variable supply and demand) than growers of commodity grains. Hence, success in IP grain production depends on the producer's ability to secure market access, capture high price premiums, maintain acceptable yields (with minimal drag), and work in a business environment driven by contracts and closer relationships. The many factors affecting IP grain producers' likelihood of success explain the high rate at which farms move into and out of specialty crop production each year. Farmers that are successful at IP grain production stick with it; those that are not return to commodity grains.

Handlers of IP grains incur indirect costs from loss of flexibility in forgoing grain mixing and the resulting underutilization of storage capacity. The magnitude of IP costs at the handling stage is influenced by the volume of grain handled, levels of purity required, handling infrastructure, and the extent of risk and risk-sharing.

With increased differentiation, grain attributes require more information management and documentation. This added transparency brings new risks and potential liability to suppliers. Depending on the type of IP grain, testing or process certification can be applied for quality assurance. Moreover, as quality becomes more critical in buyer decisions, even commodity grades for grains and oilseeds may need to be revised to reflect intrinsic qualities valued by end-users and to reward farmers capable of producing such grades. When testing is not feasible (credence attributes) or is prohibitively expensive, certification may be necessary.

The distribution of benefits from value added in the differentiated grain market has not received sufficient attention from economists outside studies on GM corn and soybean varieties. At the farm level, IP costs and yield drag may not be covered by price premiums. In this case, only high-performance farmers may earn adequate returns, and the high turnover of farmers in specialty crop production attests to this. Handlers/buyers determine their share of value captured through contracts, and technology holders can capture much of the value through intellectual property rights.

Increasing grain differentiation in the U.S. food and feed industry has raised questions about public roles. The argument for public interventions to make markets more efficient, as when USDA facilitates commodity markets, may not hold for IP grain markets. The collecting of price information for commodities is not easily extended to specialty grains, as the latter are heterogeneous, small scale, and locally concentrated. Moreover, price information can be proprietary, established through private supplier-buyer contracts. Information on specialty or IP markets has come mostly from university or private industry surveys. Likewise, USDA-approved grades for specialty grains may not be warranted since desired traits are idiosyncratic.

The public can play a supportive role in certification/testing of specialty traits and process quality, not in standard setting per se, but in providing standardized certification of privately administered testing methods.

Where direct public regulatory roles have become justified—as in the areas of public health (mad cow), safeguarding the food supply against costly disruptions (biopharmaceutical crops), and national security (bioterrorism)—resulting regulations have created new demands for identity preservation and traceability systems likely to stimulate the growth of IP grain markets. For example, the U.S. Bioterrorism and Biosecurity Act of 2002 calls for "step-back/step-forward" traceability for all food and feed moving within commercial channels. U.S. grain exports, including IP grain, are now subject to fast-changing regulatory laws that also require more traceability. Examples include the EU food traceability and labeling law and the Cartagena Protocol on Biosafety. Improving U.S. competitiveness requires government involvement in negotiations to harmonize standards, labeling, and tolerances and to fashion equivalent standards for crops, plants, and commodities entering international commerce.

REFERENCES

Ahearn, Mary, D. Banker, & MacDonald, J. (2003). *"Price and Nonprice Terms in U.S. Agricultural Contracts."* Paper presented at the AAEA annual meeting, Montreal, Canada, July 27-30.

Bard, Sharon, K., Lowell, D. Hill, Steven, L. Hofing, & Robert, K. Stewart. (2003). *Risks of Growing Value-Enhanced Corn and Soybeans in Illinois.* Savoy, IL: AEC/Centrec Consulting Group, LLC.

Bender, K. & L. Hill. 2000. *Producer Alternatives in Growing Specialty Corn and Soybeans.* Department of Agricultural and Consumer Economics, Office of Research, University of Illinois at Urbana-Champaign, AE-4732.

Bender, K., Hill, L., Wenzel, B. & Hornbaker, R. (1999). *Alternative Market Channels for Specialty Corn and Soybeans.* University of Illinois, College of Agricultural, Consumer, and Environmental Sciences, AE-4726.

Beurskens, F. (2003). "*Market Facilitation of Grain Marketing: The End-User's Perspective.*" Presentation at USDA/ERS Symposium on Product Differentiation and Market Segmentation in Grains and Oilseeds: Implications for an Industry in Transition, Jan. 27-28, Washington, DC.

Boland, M., Domine, M., Dhuyvetter, K. & Herrman, T. (1999). "*Economic Issues with Value-Enhanced Corn.*" Kansas State University Research and Extension. MF-2430.

Bullock, D., Desquilbet, S. & Nitsi, E. (2000). "The Economics of Non-GM Segregation and Identity Preservation." Paper presented at the AAEA annual meeting, Tampa, Florida, July 30-Aug. 2.

Chambers, W. & King, R. (2002). "Changing Agricultural Markets: Industrialization and Vertical Coordination in the Dry Edible Bean Industry," *Review of Agricultural Economics, 24*, 495-511.

Corn Refiners Association. (2002). Corn Annual. Washington, DC. Dahl, B. & W. Wilson. 2002. "*The Logistical Costs of Marketing Identity-Preserved Wheat*" North Dakota State University. Agribusiness and Applied Economics Report No. *495*.

Dunahay, T. (1999). "Testing May Facilitate Marketing of Value-Enhanced Crops," *Agricultural Outlook,* March. U.S. Dept. Agr., Economic Research Service.

Elbehri, A. (2005). "Biopharming and the Food System: Examining the Potential Benefits and Risks," *AgBioForum, 8(1)*, 18-25.

Elbehri, A. & Paarlberg, P. (2003). "*Price Behavior in Corn Market with Identity-Preserved Types*," Paper presented at the AAEA meetings, Montreal, July.

Farm Foundation. (2004). "Food Traceability and Assurance (TA): An Emerging Frontier Driving Developments in the Global Food System." Issue paper developed by the Farm Foundation's Traceability and Assurance Roundtable.

Fulton J., Pritchett, J. & Pederson, R. (2003). "*Contract Production and Market Coordination for Specialty Crops: The Case of Indiana.*" Paper presented at the Economic Research Service and Farm Foundation Symposium on Product Differentiation, Washington, DC, Jan. 27-28.

Giannakas, K. & Kalaitzandonakes, N. (2005). "*Economic Effects of Purity Standards in Biotech Labeling Laws.*" Paper prepared for presentation at the American Agricultural Economics Association annual meeting,

Providence, Rhode Island, July 24-27.

Ginder, R., Artz, G., Jarboe, D., Homes, H., Cashman, J. & Holden, H. (2000a). "*Output Trait Specialty Corn Production in Iowa.*" Iowa State University and Iowa Department of Agriculture. http://www.extension.iastate.edu/Pages/grain/publications/buspub/. 00gind02.*pdf.*

Ginder, R., Artz, G., Jarboe, D., Homes, H., Cashman, J. & Holden, H. (2000b). "*Output Trait Specialty Soybean Production in Iowa.*" Iowa State University and Iowa Department of Agriculture. http://www.extension.iastate.edu/Pages/grain/publications/buspub/ 00gind03.pdf.

Golan, E., Krissoff, B., Kuchler, F., Calvin, L., Nelson, K. & Price, G. (2004). *Traceability in the U.S. Food Supply: Economic Theory and Industry Studies,* AER-830. U.S. Dept. Agr., Econ. Res. Serv.

Good, D. & Bender, K. (2001). *Marketing Practices of Illinois Specialty Corn and Soybean Handlers.* University of Illinois, College of Agricultural, Consumer, and Environmental Sciences, AE-4743.

Good, D., Bender, K. & Hill, L. (2000). "*Marketing of Specialty Corn and Soybean Crops.*" University of Illinois, College of Agricultural, Consumer, and Environmental Sciences, AE-4733.

Greene, C. & Kremen, A. (2003). *U.S. Organic Farming in 2000-2001*, AIB-780. U.S. Dept. Agr., Econ. Res. Serv.

Hayenga, M. & Kalaitzandonakes, N. (1999). "*Structure and Coordination System Changes in the U.S. Biotech Seed and Value-Added Grain Market.*" Presented at the IAMA 1999 World Food and Agribusiness Congress.

Herrman, T., Boland, M. & Heishman, A. (1999). "Economic Feasibility of Wheat Segregation at Country Elevators." *Proceedings of the Annual Meeting of the National Association of Wheat Growers*, Manhattan, KS: Kansas State University, Feb.

Herrman, T., Boland, M., Agrawal, K. & Baker, S. (2002). "Use of a Simulation Model to Evaluate Wheat Segregation Strategies for Country Elevators," *Applied Transactions of the ASAE, 18(1),* 105-112.

Hicks, K., Morean, R., Johnson, D., Doner, L. & Singh, V. (2002). "*Potential New Uses for Corn Fiber,*" in Conference Proceedings: Corn Utilization and Technology Conference, 2002. Kansas City, MO.

Jensen, E. & Wilson, W. (2002). *Cooperative Marketing in Specialty Grains and Identity-Preserved Grain Markets.* Agribusiness & Applied Economics Report No. 500. North Dakota State University.

Jackson, C. & Cuppy, S. (2000). *A Producer's Guide to Specialty Grain and Oilseed Contracting.* Ohio Farm Bureau Federation. Nov.

Kalaitzandonakes, N., Maltsbarger, R. & Barnes, J. (2001). "Global Identity Preservation Costs in Agricultural Supply Chains," *Canadian Journal of Agricultural Economics, 49*, 605-615.

Lin, W., Chambers, W. & Harwood, J. (2000). "Biotechnology: U.S. Grain Handlers Look Ahead," *Agricultural Outlook*, AGO-270, ERS-USDA.

Maltsbarger, R. & Kalaitzandonakes. N. (2000), "Direct and Hidden Costs in Identity Preserved Supply Chains," *AgBioForum, 3(4)*, 236-42.

Marks, Eric, A. & Mark, J. (2003). Werrell. *Executive's Guide to Web Services*. John Wiley and Sons, Inc., Hoboken, NJ.

Moss, C. B., Schmitz, T. G. & Schmitz, A. (2002). *"Differentiated GMOs and Non-GMO's in a Marketing Channel."* Paper presented at the 6th International ICABR Conference, Ravelo, Italy.

Phillips, P. & Smyth, S. (2003). *"Identity Preservation in Marketing Systems in Canada: Developments in Wheat and Canola Sectors."* Presentation at USDA/ERS Symposium on Product Differentiation and Market Segmentation in Grains and Oilseeds: Implications for an Industry in Transition, Jan. 27-28, Washington, DC.

Qasmi, B., Van der Sluis, E. & Wilhelm, C. (2004). *"Cost of Segregating Non-Transgenic Grains at Country Elevators in South Dakota."* Paper presented at the Western Agricultural Economics Association Meeting, Hawaii.

Reichert, H. & Vachal, K. (2000). *"Identity-Preserved Grain—Logistical Overview,"* Draft paper, Upper Great Plains Transportation Institute, North Dakota State University.

Shipman, D. R. (2003). *"Public Role and Oilseed Marketing Facilitation: USDA Perspective."* Presentation at USDA/ERS Symposium on Product Differentiation and Market Segmentation in Grains and Oilseeds: Implications for an Industry in Transition, Jan. 27-28, Washington, DC.

Smyth, S. & Phillips, P. (2003). "Product Differentiation Alternatives: Identity Preservation, Segregation, and Traceability." *AgBioForum, 5(2)*, 30-42.

Sonka, S. (2003). *"Forces Driving Industrialization of Agriculture: Implications for the Grain Industry in the United States."* Presentation at USDA/ERS Symposium on Product Differentiation and Market Segmentation in Grains and Oilseeds: Implications for an Industry in Transition, Jan. 27-28, Washington, DC.

Sporleder, T. & Schmidt, P. (2003). *"Differentiation Within the Grain and Oilseeds Sectors: The Evolution and Reengineering of Supply Chains Increased."* Paper presented at USDA/ERS Symposium on Product Differentiation and Market Segmentation in Grains and Oilseeds:

Implications for an Industry in Transition, Jan. 27-28, Washington, DC.
Stewart, R. (2003). "*2001-2002 Value-Enhanced Grains Quality Report: Producer Survey Results.*" Presentation at USDA/ERS Symposium on Product Differentiation and Market Segmentation in Grains and Oilseeds: Implications for an Industry in Transition, Jan. 27-28, Washington, DC.
Sykuta, M. & Parcell, J. (2002). "*Contract Structure and Design in Identity Preserved Soybean Production.*" Contracting and Organizations Research Institute, University of Missouri, CORI Working Paper No. 02-01.
Vanderburg, J., Fulton, J., Dooley, F. & Preckel, P. (2003). "*Impact of Identity Preservation of Non-GMO Crops on the Grain Market System.*" Staff Paper #00-03. Purdue University.
Wilson, W., DeVuyst, E., Koo, W., Taylor, R. & Dahl, B. (2005). *Welfare Implications of Introducing Biotech Traits in a Market with Segments and Segregation Costs: The Case for Roundup Ready® Wheat – Summary*. Agribusiness & Applied Economics Report No. 566-S, North Dakota State University, Oct.

End Notes

[1] In seed production, individual lots of seeds are catalogued by variety for each stage from breeder to seed registration and certification. Involved parties, including growers and seed companies, keep all records of production, field inspections, inventories, transportation moves, cleaning, certification tags, etc., for a minimum of 3 years. Compliance is verified either by third-party auditors or by crop improvement personnel from seed companies.

[2] Internet-based service providers record the performance of the critical steps and the results of the trait tests from remote producer, shipper, buyer, and laboratory locations. The Internet also provides a cheap way to process and save the recorded information and to make it immediately accessible to all market participants.

[3] For example, Genetic ID, an Iowa- based certification firm specializing in identity-preserved food and feed products, serves as go-between for grains and ingredient suppliers (such as Cargill, ADM, Kerry, or Cerestar) and food manufacturers and retailers (such as Sainsbury, Nestle, and Safeway). Genetic ID offers a proprietary package service *CertID* that includes testing, validation, inspection, documentation, and certification with a proprietary seal. Such services certify that shipments are non-GM, organic, or have specific traits, depending on the client.

[4] USDA's Agricultural Resource Management Survey (ARMS) for 2001 shows that the additional expenses of storage, segregation, and transportation are substantial for specialty corn, while cleaning of planters and combines is negligible.

[5] For identity verification, the costs include processes to review and record details of delivered loads and supporting certification, and costs to validate claims about IP (Maltsbarger and Kalaitzandonakes, 2000).

[6] Herrman et al. (2002) report that delays associated with segregating wheat during harvest accounted for 15.8 to 27.5 percent of total segregation costs.

[7] Oil content that is lower by 1 percent might reduce price premiums paid to high-oil corn

producers; 1-percent biotech content in a non-GM grain shipment could cause rejection—a much bigger penalty for noncompliance, particularly for exporters.

[8] Testing for GM presence can be done through detection of proteins associated with transgenes, or detection of the transgene itself in DNA. The relative ease or complexity (and hence cost) of a test depends on the nature of the product (whole-grain, semi-processed, or processed) and the amount of target protein or DNA that can be detected. The higher the amount of protein and the more accurate the measurement technology, the lower the probability of false- positive results.

[9] The incident involving ProdiGene, Inc., a biotechnology firm, illustrates the kind of risks facing the food industry. In Nebraska during the 2002 season, APHIS inspectors discovered "pharmaceutical" corn from the previous season growing in the midst of a soybean field. As a result, both the harvested soybeans (500 bushels) and the entire soybean load of 500,000 bushels in the local elevator were quarantined and ProdiGene was fined.

[10] In some cases, specialty grains have become "commoditized" (white corn, white wheat) and spot markets endure.

[11] Fulton et al. (2003) found that for Indiana growers of specialty grain, the dominant reason for entering into production contracts was additional rev- enue (92 percent of respondents), while a third indicated market access as a reason. About 28 and 21 percent of respondents, respectively, cited access to seed and reduction of risks as important.

[12] The Cartagena Protocol on Biosafety was adopted on 29 January 2000, signed by 107 parties, and by September 2003 was ratified by 50 countries, the minimum required for the Protocol to enter into force. Countries that ratified the Protocol became Parties to the Protocol and are required to comply with and implement all of its provisions. Countries that have not signed but that export Living Modified Organisms (LMOs) to member countries are encouraged to comply with the Protocol's provisions implemented in the importing country.

INDEX

A

access, 56, 57, 63, 73, 79
accountability, 62
acid, 44, 45, 49
adaptation, 51
adaptations, 40
adults, 3, 4, 6, 7, 9, 10, 12, 16, 17, 23, 25, 26
age, 7, 8, 9, 10, 19, 20, 25, 26
agencies, 51
Agricultural Research Service, 7, 30
amylase, 46
ARS, 7
assessment, 61
audit, 40, 51
audits, 47
authentication, 40, 46, 62
autonomy, 68
aversion, 45, 66
awareness, 7

B

bargaining, 68
base, 28, 58, 61, 78
benefits, 6, 23, 63, 69, 71, 73
biotechnology, 41, 44, 45, 51, 79
bioterrorism, 74
Blacks, 3, 11, 12, 13, 16, 17, 27
business environment, 73
businesses, 51
buyer, 40, 50, 58, 63, 65, 67, 73, 74, 78
buyers, 38, 39, 44, 52, 57, 58, 60, 63, 65, 66, 68, 70, 72, 73

C

caloric intake, 5, 9, 10, 11, 12, 13, 14, 15, 16, 18
calorie, 3, 8, 9, 12, 14, 16
campaigns, 5
cancer, 5
carbohydrate, 42, 72
carbohydrates, 42
case studies, 41
cash, 63
CDC, 7, 28
cellulose, 46
certification, 44, 50, 51, 52, 62, 70, 73, 74, 78
children, 3, 7, 9, 11, 12, 16, 17, 19, 21, 27
cholesterol, 19
cleaning, 49, 78
combined effect, 25
commerce, 52, 74
commercial, 71, 74
commodity, 38, 39, 40, 41, 45, 46, 47, 48, 51, 52, 56, 57, 58, 62, 63, 66, 69, 70, 71, 72, 73
commodity markets, 40, 46, 47, 70, 73
community, 5

compensation, 63
competition, 47, 63
competitiveness, 74
competitors, 63
complexity, 72, 79
compliance, 39, 54
composition, 56, 61
computing, 50
conditional mean, 35, 36
configuration, 54
confinement, 48, 50, 61
consumer demand, 72
consumer goods, 44
consumers, vii, 1, 2, 6, 9, 13, 16, 17, 19, 25, 44, 71, 72
consumption, iv, vii, 1, 2, 3, 4, 5, 6, 7, 8, 9, 10, 11, 12, 13, 14, 15, 16, 18, 19, 20, 21, 23, 25, 26, 27, 34
containers, 50
contamination, 39, 48, 53, 56, 57, 60, 71, 72
Continuing Survey of Food Intakes by Individuals, 1, 4, 6, 7, 29, 30
coordination, 37, 38, 39, 40, 53, 62
coronary heart disease, 5
correlation, 24
correlations, 32, 34
cost, 20, 38, 39, 40, 41, 45, 47, 48, 49, 52, 53, 54, 58, 61, 62, 63, 72, 79
cost structures, 48
costs of production, 44
cotton, 62
covering, 41
critical period, 61
crop, 39, 42, 45, 48, 49, 56, 57, 58, 59, 60, 64, 65, 67, 68, 70, 72, 73, 78
crop insurance, 70
crop production, 59, 73
crops, iv, vii, 37, 38, 39, 40, 41, 42, 44, 45, 46, 48, 49, 50, 51, 52, 53, 55, 58, 60, 61, 62, 65, 66, 67, 68, 69, 70, 71, 72, 73, 74
CSFII, 4, 6, 7, 8, 10, 12, 14, 17, 18, 19, 20, 23
customers, 52, 54, 63, 70
cycles, 68

D

deficiency, 6, 16
demographic characteristics, vii, 1, 3, 4, 6, 7
demographic data, 20
Department of Agriculture, vii, 1, 2, 28, 30, 31, 37, 38, 76
Department of Health and Human Services, vii, 2, 28, 30, 31
dependent variable, 19, 25, 32, 35
derivatives, 34
destruction, 61
detection, 79
deviation, 36
DHKS, 7, 19, 22, 23
diet, vii, 2, 3, 4, 5, 6, 7, 8, 19, 21, 22, 24, 26, 27
Diet and Health Knowledge Survey, 4, 7, 22
dietary fiber, 42
Dietary Guidelines, vii, 1, 2, 3, 5, 6, 8, 9, 10, 11, 14, 15, 30, 31
dietary intake, 6, 7, 19
differentiated products, 72
directives, 51
disaster, 57
disclosure, 69
discrete variable, 36
diseases, 1, 5
distribution, 34, 35, 73
diversity, 47, 66
DNA, 79
domestic markets, 70
drying, 53

E

e-commerce, 52
economic incentives, 41, 52
economics, 39, 47, 59
economies of scale, 20, 54
education, 14, 20, 22, 25, 34
educational attainment, 3
endogeneity, 34
endosperm, 4
end-users, 45, 50, 52, 55, 73

energy, 8, 16, 18, 19
engineering, 72
environment, vii, 38, 51, 71, 73
enzyme, 72
enzymes, 46
equipment, 47, 48, 53, 59, 61, 64
ethnicity, 3, 11, 12, 13, 17, 20, 25
EU, 74
Europe, 60
European Union, 51, 71
evidence, vii, 1, 2, 5, 65
exclusion, 34
exercise, 3, 20, 35
exporters, 79
exports, 60, 74
externalities, 69

F

farmers, 41, 44, 52, 53, 58, 59, 60, 64, 65, 67, 68, 72, 73
farms, 45, 51, 59, 68, 72, 73
fat, 7, 19, 21, 22, 23
fat intake, 7
FDA, 31, 71
Federal Government, 6
fertility, 56
fertilizers, 51
fiber, 5, 19, 21, 23, 42, 46, 72
fiber content, 23
fibrosis, 47
financial, 58, 63
fixed costs, 53
flexibility, 38, 49, 54, 73
flour, 4, 19, 48
fluctuations, 54, 59
food, 1, 2, 3, 4, 5, 6, 7, 8, 10, 17, 18, 19, 20, 23, 24, 26, 27, 32, 37, 40, 41, 42, 44, 45, 46, 48, 49, 50, 51, 53, 57, 58, 61, 62, 63, 66, 67, 69, 71, 72, 73, 74, 78, 79
food industry, 5, 6, 72
food intake, 4, 7, 19
food products, 44, 62
food safety, 7, 37, 51, 52, 57, 63, 72
force, 79
formation, 6

formula, 66
free trade, 71
fruits, 5, 8
functional food, 46

G

genetic marker, 70
Georgia, 2
globalization, 69
government intervention, 69, 70
government policy, 37
grades, 39, 40, 51, 53, 62, 69, 70, 73, 74
grading, 51, 70
grants, 65
growth, 63, 71, 74
guidance, 38
guidelines, 2, 3, 7, 8

H

harmonization, 70, 71
harvesting, 53, 64, 70
Hawaii, 77
health, vii, 2, 3, 4, 5, 6, 7, 23, 30, 42, 61, 71, 72, 74
Health and Human Services, vii, 2, 28, 30, 31
health promotion, 6
heart disease, vii, 2, 5
height, 71
hepatitis, 49
herbicide, 44
heterogeneity, 36
high school, 15, 22
high school diploma, 15
higher education, 3, 14
Hispanics, 3, 11, 12, 13, 16, 17, 27
homogeneity, 41
household income, 3, 16, 20, 25
hybrid, 65

I

identification, 34, 50
identity, vii, 37, 38, 39, 40, 41, 45, 47, 48, 50, 51, 52, 53, 61, 71, 72, 74, 78

idiosyncratic, 40, 74
imbalances, 68
impacts, 61
impurities, 54, 60
income, 3, 13, 14, 16, 20, 21, 25, 72
independence, 34, 36, 65
indirect costs, 54, 73
indirect effect, 25
individuals, 7, 14, 23, 25, 27
industrial processing, 41, 72
industrialized countries, 42
industries, 61
industry, 5, 6, 37, 40, 41, 44, 46, 47, 50, 52, 70, 72, 73, 79
infrastructure, 51, 53, 54, 71, 73
ingredients, 23, 40, 44, 46, 62
inspections, 47, 50, 61, 65, 78
inspectors, 79
integrity, 51
intellectual property, 62, 63, 64, 66, 73
intellectual property rights, 64, 66, 73
Internet, 46, 52, 72, 78
intervention, 2, 6, 69, 70, 71
intervention strategies, 2, 6
investment, 59, 64
investments, 45, 56, 57, 58, 59, 63
iodine, 50
Iowa, 70, 76, 78
isoflavone, 43
isolation, 58, 61
Italy, 77

J

Japan, 29, 60

L

labeling, 51, 57, 71, 74
laws, 74
lead, 39, 45, 52, 66
leadership, 38
light, 21, 23
liquidity, 47, 69
livestock, 64, 68
logistics, 46, 72

LTD, 42
lysine, 43

M

magnitude, 73
management, 37, 38, 40, 41, 47, 51, 53, 54, 58, 60, 61, 62, 63, 64, 65, 68, 73
manufacturing, 20, 40, 42, 62
manure, 45, 72
marginal distribution, 34
market access, 73, 79
market segment, vii, 38, 41
market structure, 47
marketability, 50
marketing, 5, 20, 37, 39, 40, 41, 45, 46, 48, 49, 50, 51, 52, 54, 63, 64, 71, 72
materials, 40, 62
matrix, 32, 34, 35
measurement, 8, 41, 44, 50, 51, 56, 70, 79
measurements, 70
Missouri, 78
mixing, 73
model specification, 18
modeling, 6
models, 32, 34
modification, 60
modifications, 48
monitoring, 6, 46, 60
motivation, 36

N

National Health and Nutrition Examination Survey, 7, 28
National Research Council, 5
national security, 74
Nationwide Food Consumption Survey, 19
natural disaster, 57
negotiating, 68
NHANES, 7, 28
niche market, 42, 62
NIR, 50
normal distribution, 35
nutrient, 7, 19
nutrients, 19, 23

nutrition, 6, 20, 23, 25, 32, 34
nutrition labels, 20, 25

O

oil, 41, 43, 47, 48, 49, 50, 53, 54, 55, 61, 64, 66, 67, 68, 78
oilseed, 63, 68
omission, 36
operations, 56
opportunities, 41, 46, 52, 53
opportunity costs, 53
overproduction, 56
oversight, 71
overweight, vii, 2
ownership, 65, 67

P

Pacific, 22
participants, 78
partition, 34, 35
pasta, 21, 22
penalties, 60, 65
performance, 40, 57, 58, 59, 62, 73, 78
pests, 45
pharmaceutical, 37, 48, 58, 61, 73, 79
pharmaceuticals, 61, 62, 71
phosphorus, 45, 72
physical activity, 10
plants, 54, 61, 62, 74
Plato, 38
policy, 6, 37, 47
policy issues, 6
pollen, 53
pollination, 61
pollution, 45, 72
poor performance, 65
preservation, vii, 37, 38, 39, 41, 47, 48, 49, 51, 52, 53, 63, 64, 71, 72, 74
price changes, 66
private firms, 70
probability, 25, 26, 32, 79
probability density function, 32
procurement, 53
producers, 39, 40, 44, 45, 48, 51, 52, 58, 59, 60, 65, 66, 68, 69, 70, 71, 73, 79
product attributes, 40, 44, 62
product market, 44
production costs, 45
profit, 6
project, 38, 41
property rights, 64, 66, 73
protection, 48, 62, 66
proteins, 79
public health, 5, 74
purity, vii, 38, 39, 48, 50, 53, 54, 58, 60, 65, 72, 73

Q

quality assurance, 37, 38, 41, 48, 50, 73
quality control, 40, 61, 63, 64, 70
quality standards, 39, 69

R

race, 3, 11, 12, 13, 17, 20, 22, 24, 25, 26
raw materials, 40
reading, 23
real-time basis, 69
recall, 7, 19, 52
recognition, 71
recommendations, iv, 1, 2, 5, 6, 7, 8, 10, 12, 16
regression, 6, 23
regression analysis, 6, 23
regulations, 37, 49, 51, 60, 61, 71, 74
regulatory oversight, 71
rejection, 39, 40, 56, 58, 60, 79
reliability, 63
rent, 64
requirements, 37, 39, 42, 44, 48, 50, 52, 53, 54, 61, 63, 64, 68, 71
researchers, 19
resistance, 45
response, 42, 53, 54, 69
restaurants, 28
restrictions, 32, 34
retail, 39, 45, 72
rights, iv, 64, 65, 66, 73
rings, 62, 73

risk, vii, 1, 2, 5, 37, 38, 39, 40, 41, 47, 53, 56, 57, 58, 59, 60, 61, 62, 63, 66, 67, 68, 73
risk assessment, 61
risk aversion, 66
risk factors, 39, 47
risk management, 37, 38, 40, 41, 61
risks, 39, 40, 52, 54, 58, 59, 61, 62, 63, 65, 67, 72, 73, 79
root, 36
rules, 63

S

safety, 7, 37, 44, 51, 52, 57, 63, 71, 72
school, 6, 15, 16, 22, 24
science, 6
scope, 58, 72
security, 71, 74
seed, 40, 47, 49, 50, 52, 53, 61, 64, 65, 67, 70, 78, 79
segregation, vii, 37, 38, 39, 40, 41, 44, 45, 47, 52, 53, 54, 55, 60, 69, 72, 73, 78
sellers, 38, 63, 70
service organizations, 52
service provider, 52, 78
services, iv, 40, 52, 70, 78
showing, 62
signals, 53, 72
simulation, 34
smoking, 20, 22
social costs, 69
sodium, 19
software, 46, 52
South Dakota, 77
soybeans, 41, 42, 43, 44, 45, 48, 49, 50, 53, 55, 60, 62, 65, 66, 68, 79
Spain, 29
specialty crop, vii, 38, 39, 40, 41, 42, 48, 49, 55, 59, 70, 72, 73
specifications, 23, 60, 65, 70
spending, 19
spot market, 62, 65, 79
standard deviation, 32, 36
starch, 42, 44, 45, 61, 72
statistics, 6, 21

stomach, 5
storage, 46, 53, 61, 64, 73, 78
structural changes, 72
structure, 20, 27, 37, 39, 47, 52, 58, 61, 62, 68, 72
subgroups, 16
substitutes, 39
substitution, 68
sucrose, 43, 49, 68
sulfonylurea, 62
supplier, 40, 46, 52, 74
suppliers, 39, 40, 45, 57, 62, 63, 64, 65, 66, 70, 73, 78
supply chain, 39, 42, 44, 45, 48, 49, 50, 51, 52, 53, 54, 61, 62, 63, 71
supply curve, 45
survey, 1, 2, 4, 6, 7, 18, 19, 25, 47, 59, 64, 66, 70

T

technological advances, 69
technologies, 44, 46, 63
technology, 6, 53, 57, 60, 63, 65, 66, 70, 73, 79
temperature, 43, 66
terrorist attacks, 71
testing, 39, 40, 44, 47, 50, 51, 52, 53, 56, 58, 60, 64, 65, 69, 70, 73, 74, 78
theft, 57
tofu, 42, 43, 49, 50, 53
trade, iv, 70, 71
trade agreement, 71
training, 61
traits, 40, 41, 43, 44, 45, 47, 48, 49, 51, 61, 69, 70, 72, 74, 78
transaction costs, 40, 46, 52, 62, 63, 66, 72
transactions, 40, 62
transgene, 79
transparency, 40, 62, 73
transport, 53
transportation, 46, 53, 54, 72, 73, 78
treatment, 32
turnover, 73
typology, 56

U

UK, 44
United, iv, 1, 7, 16, 20, 27, 30, 37, 41, 43, 58, 77
United States, iv, 1, 7, 16, 20, 27, 30, 37, 41, 43, 58, 77
universities, 6
urbanization, 35
USA, 29
USDA, 1, 2, 4, 5, 6, 7, 17, 19, 38, 40, 51, 61, 66, 69, 70, 71, 73, 75, 77, 78

V

validation, 62, 78

variables, 19, 20, 21, 23, 25, 26, 32, 34, 35, 36
variations, 19
varieties, 40, 41, 42, 44, 45, 48, 53, 60, 62, 63, 64, 65, 66, 70, 73
vegetables, 5, 8

W

warehouses, 51
Washington, 29, 75, 77, 78
workers, 61

Y

yield, 40, 41, 44, 48, 56, 58, 59, 63, 66, 67, 69, 73